Mr. Secretary

Mr. Secretary—
From the Potato Fields to the White House
By Wilbert Bryant,
Colonel U.S. Army (Retired)

Copyright © 2026
by Julie Wakeman-Linn
All Rights Reserved

Collaborative Editor
Julie Wakeman-Linn

Publisher
New Bay Books
Fairhaven, Maryland
NewBayBooks@gmail.com

Cover design and photo collage by Suzanne Shelden
Interior design by Suzanne Shelden
Shelden Studios
Prince Frederick, Maryland
sheldenstudios@comcast.net

A note on type:
Cover and section heads and text is Garamond Premier Pro.

Library of Congress
Cataloging-in-Publication Data
ISBN 979-8-9882998-9-9
Printed in the United States of America
First Edition

Mr. Secretary—From the Potato Fields to the White House

to my mother
Claudia M. Bryant

"Colonel Wilbert Bryant is the epitome of grit. Just look up 'grit' in Webster's Dictionary and you just might see his portrait. Mr. Secretary is an inspirational story for all walks of life and serves as a blueprint for a life modeled on resilience, courage, hope, and perseverance. This poor Black child and his four brothers were shielded from the limitations of the segregated south. Mom 'Claude' raised her sons single-handedly, preaching their dreams were unlimited with the cornerstone of education. Wilbert and his brothers heeded their Mom's directive and each achieved their own 'rags to riches' stories in the military, medicine, dentistry, state politics, and the White House.

"You will cheer Wilbert on as he overcomes the odds with his true grit based on respect, hard-work, hope and courage. His memoir is promised to make you smile and shed some tears. Wilbert's journey of the soul just might encourage you to write your own story!"

—Barbara J. Wickham served as Aide-de-Camp and Chief of Protocol for Wilbert Bryant's oldest brother Major General Alvin Bryant, Commanding General of the 310th TAACOM, Fort Belvoir, VA (He was the highest ranking African American Army Reserve Officer in 1986). Barbara is a Desert Storm veteran, retiring as a senior officer after 27 years of service in the U.S. Army. She also wrote for *The Miami Times* and *The Denver Post*.

CONTENTS

Prologue 1
Chapter One: Lessons Learned from Claude and My Brothers 5
Chapter Two: Coaches and Mentors 23
Chapter Three: Straight Outta Goulds 37
Chapter Four: Romancing Emily 49
Chapter Five: Army Green 67
Chapter Six: German Immersion 83
Chapter Seven: Vietnam: Leadership Under Fire 91
Chapter Eight: Korean Year of Losses and Fixing Problems 113
Chapter Nine: Commandant of Cadets Meets
    The Black Panthers 129
Chapter Ten: England and the VW Bug that Crossed the Alps ... 147
Chapter Eleven: The Pentagon and Helpful Neighbors 167
Chapter Twelve: California and Command 179
Chapter Thirteen : War College 199
Chapter Fourteen: High School Reunions and a
    High School Star 213
Chapter Fifteen: Shaking up Virginia Union University
    and Virginia Republicans 225
Chapter Sixteen: A Parent's Worst Nightmare 241
Chapter Seventeen: Working for Governors 255
Chapter Eighteen: Mr. Secretary Goes to the White House 269
Chapter Nineteen: Strengthening HBCUs and
    the Peace Corps 287
Epilogue 303
Acknowledgments 305
Photo Captions 307

Mr. Secretary

From the Potato Fields to the White House

By Wilbert Bryant
Colonel, U.S. Army (Retired)

With Julie Wakeman-Linn

Prologue

One summer at twilight in the mid-1940s my brothers and I were playing in front of our grandfather's ramshackle house in our rural town of Goulds, Florida. My next younger brother and best friend, Henry, and I shot marbles across the dirt in front of the house. I wanted to win his cat's eye. My older two brothers, Alvin and Willie, tossed around an old baseball they'd found. Bugs buzzed. A slight breeze cooled us, stirring the leaves of the mango tree. My mother was still at work and my grandma was inside, probably cooking and caring for my baby brother.

"Boys, Boys!" My grandpa ran down Old Cutler Road from the direction of our nearest neighbor, the next little four-acre farm. I started to laugh at his funny stride, but his face looked strange. His mouth open as he panted, he kept glancing over his shoulder. Alvin, the oldest of us, dropped his ball and raced to him.

"Get all your brothers inside the house. Now!" Grandpa yelled, waving his arms toward the front door. He never yelled.

"Come on. Go!" Alvin, the oldest of us, hurried us to the porch.

"You gotta be inside. If they see you, they will shoot you." Grandpa's hand gripped his chest. He struggled to pull in air even as he kept running to us.

We obeyed. He seemed so stern this twilight, when he usually smiled at us. "All of you get down on the floor, lie flat." He yanked down the paper roller shades, cursing when one flapped up and had to be pulled again.

I lay down on the scuffed rough wood, trying to avoid splinters, near the front wall of the house, closest to the porch and the yard and the road.

"Alvin, turn off the lamp. They mustn't see you." Alvin belly-crawled across the floor to our kerosene lantern and turned off the flame.

Grandpa dropped on one knee, steadied himself with his hand, and flopped his body to the floor, across the front door. He flattened his body against the floor. I reached and held Henry's hand.

Turning my head toward Alvin, I whispered, "What is it?"

"KKK. They're coming with their guns. Hush." Grandpa whispered. He was always ready to laugh with us, but not this evening. "Keep your heads down."

KKK. I'd heard the whispers of these white men, their ropes, white sheets and guns—catching Blacks, beating them, dragging them, killing them.

I heard the rumbling of a big pickup truck, getting closer, driving down our road, past our house. I heard a shot fired from a gun. Squeezing my eyes shut, my cheek on the floor, I tried to flatten myself even more. The men in the truck yelled now, nasty things. I looked to Alvin, three years older than me, and started to say, "What will..."

"Shut up. Don't talk." Alvin mouthed the words without making a sound.

I wanted to peek out a window and see for myself these dangerous white men, but I was too afraid, frozen to the floor, unable to move.

Later I would ask Alvin and Willie how we could survive, but I knew deep inside, we brothers had to find a way out of Goulds' poverty and violence to make something of our lives.

Chapter One

Lessons Learned from Claude and My Brothers

I was born and raised in Goulds, a segregated town of less than 2000 people, 26 miles south of Miami. In the early 1940s, even though we were segregated, we were a genuine community because we knew all the families along our rural road and cared for each other.

At the age of 10, my mother had suffered frostbite. In the 1920s, her community in Northern Florida had almost no healthcare for Blacks. When she finally got to a doctor, she had to have her leg amputated above her knee or gangrene would have killed her. She got her wooden leg at age 11. Her grandmother, Mary Hill Williams, sent for her and raised her in Goulds when her mother Mattie died.

We called our mother Claude, not Mama. Even though her name was Claudia Mae Bryant, everybody called her Claude. She, a single parent, worked every day as a domestic even though she was disabled.

My mother began having her seven children when she was 16 or 17 years old.

Alvin "Ray" Bryant was the oldest, born in February 1937. My mother put him in charge of us, making him our leader. He was always the tallest one. Willie Lee "Pat" Bryant, the second child, was born in October 1938 and he was Alvin's quiet wingman. I, the third child, was born in July 1940. Often I felt stuck in the middle with lots of chores. Henry Louis Bryant, fun-loving and smiley, was born in September 1941.

Another younger brother, Sylvester Bobby Bryant was born in May 1944. My youngest brother, Claude Larry Bryant, was born in April 1951. I had one sister, Brenda Lee Bryant, who was a surprise to us boys, born in March 1954.

With Alvin and Willie being so close, Henry and I formed a very close bond. He was my best friend. We played everything together—marbles, hopscotch, sticks, ping-pong—always competitively. Henry and I played ping-pong with sandpaper paddles against each other and then in our summer school program. Many times I'd win my age group and he'd win his. We did the washing and ironing together; we studied together. Although he was more outgoing than me with many friends, he was my best friend.

We, the first five, lived in our elderly great-grandparents' house, my mother's mother's parents. My mother's father, Preston Bryant and her stepmother and their son, also lived with us for a time until they could move to a place of their own. All these people lived in an old wooden decrepit house.

The house only had two bedrooms. All eleven of us slept head to toe, some on a floor mattress. During World War II, a very tough time, everything went to the war effort; food was scarce, there was no money and no jobs. We received ration coupons that allowed us

to get basics, like kerosene fuel, milk, flour, bread and lard. We ate reasonably well because our great-grandfather and our great-grandmother had all this on their small farm: chickens, an avocado tree, a mango tree, and a grapefruit tree. They had a cow for a short time to give us fresh milk.

This old house had no electricity, no running water and an outhouse. Daily I primed the handle pump on the back porch and hauled fresh water into the kitchen. A two-burner kerosene oil heater in the middle of the living room warmed the whole house. The older family members plugged the holes in the walls with newspaper to try to keep the heat in, although the temperature outside never got below 60 degrees. My mother and great grandmother cooked on a wood stove in the kitchen. My great grandfather and my two older brothers would go into the woods behind us and cut firewood, bring it home, and stack it in the wood box. I brought wood from the outside box to the kitchen wood box. With Alvin and Willie off at school activities, often I got stuck with looking out for the younger brothers and doing a lot of chores.

Many times my mother would bring leftover food from the White homes where she worked in Miami Beach or South Miami. When she got home, she'd tell us about their homes. She said, "We want to get us a nice home like I work in. You all have to help me build a new one." Our decrepit house increased our desire to earn money to help our mother.

When I was seven, cousin Willie James Thomas visited us and showed us photos of him in his Army uniform. He told us boys stories of all the places he'd been in the Army in the war. He said if we got good grades, we could perhaps earn academic or sports scholarships for college. I wanted to be like him, see the world, wear a uniform, and get out of Goulds.

We knew who our father was, but he was married to another woman. We were all born out of wedlock. We knew who his wife was, and he had one daughter with her. When I was 8, an age when four of us could work, we worked for our father who was the contractor foreman for H.L. Cox and Sons, a potato company. With his two trucks, he'd collect a crew of workers to pick up the potatoes after a tractor plowed each row in the potato field. I remember thinking they were beautiful white potatoes. My oldest brother, Alvin, and my brother Henry, picked together, and I teamed with Willie. Willie said in his quiet voice, "You gotta keep up with me." On our knees, we crawled, picking up the potatoes and putting them in a wood box. We'd drag our box along our row, which was a section about 20 feet wide and the length of the entire field, maybe 300 yards long. I rushed to keep up with Willie so we could move the box at the same time.

We earned five cents per box. Our goal was to get a minimum of 100 boxes to get five dollars; my mother only made five dollars a day plus 50 cents carfare. If we really picked fast, our brother teams could make twice in a day what Claude made.

In the field, a crew of about 40 people, all Blacks, wore straw hats to shield us from the sun. Dust covered our faces. Our mother prepared us one lunch pail of cooked black-eyed peas, rice, and some kind of meat, neck bones or a sausage and cornbread. She'd send four spoons, five when Bobby was out there. We worked in the fields during the warm weather in April and probably May or June.

Alvin, our ever-ambitious leader, wanted us to finish our section, then go help some of the others who lagged behind us as a way to get extra boxes to increase our daily totals. I loved to eat, but I didn't want to work too hard. I didn't like working in the field, but Alvin pushed us to work as hard as he did. Willie and I competed with

Alvin and Henry, but if we picked 110 to 120 boxes, they'd pick 125 to 135. We brothers competed in everything from roller skating time to ping-pong games to picking potatoes.

My fifth brother, young Bobby, also helped when he was old enough. His first task was to run and get boxes for us. Bobby's other job was to go to the checker while she counted our boxes and she gave him our tickets for each one. In the evening he turned those tickets over to my oldest brother. H.L. Cox and Sons made the payment envelope out to "Alvin Bryant." I wanted my name on my own envelope, but they wouldn't. I was working—I wanted to give my mother the money I had earned.

We worked from about 7:30 in the morning until about five or six in the evening.

Sometimes we would stay out of school twice a week to pick up the potatoes. Even though we stayed out of school in harvest season, we'd go back and catch up in our classes.

Back in school, I visited each teacher to receive the information I missed. At home, I completed all the missed assignments, and my two older brothers checked to ensure my work was complete.

My mother wanted us to study hard and get a good education, so we didn't have to do field work after we left school. She told us the white people she worked for in South Miami and Miami Beach sent their children to college. She said, "You can wear a coat and tie just like Mr. Johnson. You can do all those things." Claude didn't get to be educated, but she believed it could deliver us to a better life. My mother would get up two hours early every morning, cook breakfast for us and make sure we washed up and ate our breakfast before she left so we wouldn't go to school hungry. She wanted us to concentrate fully on our studies. Alvin, my oldest brother, was in charge to make sure we were ready on time.

Claude encouraged us to read even before we began our schooling. My childhood was long before the Brown vs the Board of Education decision in 1954. We believed education was so important, but Goulds lacked resources. We didn't have a library in our town, although we had a small school library. We had to put our name on a list to get a book to read and we had to return it in three days. Sometimes we had access to *Life Magazine*, *National Geographic*, and *Ebony* magazine.

Our teachers' dedication provided another compensation for the lack of resources. They knew we Bryants were poor, but they treated us fairly. We were all good students with good manners. My mother insisted we never talk back to adults. Social promotion was not part of the school policy. If a student didn't earn the grades to pass, then he or she was kept back in the grade. Some students dropped out of school when they were held back; I didn't intend that to happen to me.

In third grade, I was a spunky kid. My mother wanted me to learn multiplication tables from the back cover of a blue tablet. She told me to learn those multiplication tables by the end of the week. I tried hard the whole week. She had Alvin and Willie test me. I thought, *Claude, I don't want you to let them tell me what to do.* I told my brothers I knew the times tables.

Alvin was ready to quiz me. He said, "Give me the tablet. Let's start in the nines."

"No." I said. "That's not how I learned it. I started on the ones. It has to be in sequence."

"You don't know the timetables!" he said.

I argued I did know them!

We fussed. I threw a punch at him. When my mother came home and he told her, she told me to go cut a switch, a thin tree branch. She wore me out, her way of giving a spanking. She said, "Monday, when I come home, I want you to know those timetables front to back, middle to back to front."

This time I learned them; I did not want to get that spanking with a switch. Pretty soon, I learned how to be smart and manage switches. I'd crack them in the middle and they would break after three swats. The next year I quizzed Louis when it was his time to learn the multiplication tables as well.

Throughout elementary school and into junior high, we boys kept growing as all children do. Facing the challenges of feeding and clothing all of us, my mother took our father to court to try to get support for us. He, a prominent businessman in our community, disowned all six of us, his sons. His action just increased our motivation to earn money to build our own home ourselves. My brothers and I doubled down and worked harder.

One day a lady from the county showed up at our home because the officials had heard about this mother in Goulds with six children and no father to help. The county lady offered my mother used clothes. My mother sent off the two youngest boys to another room. My two older brothers and I were sitting with her. My mother told the county lady at the time, "I'll let you know what we decide." As soon as she walked out of the door, my two older brothers told my mother, "We do not want any hand-me-down clothes. We can go out and work and save and pay for our own clothes."

"Okay," she said, "that's the way we'll do it." She taught us self-reliance early on. She was always thinking and planning ahead.

My mother always told us that if anything ever happened to her, go get her artificial leg. She packed a sock in her leg, a hollow spot where she kept all the cash we had earned. Every night, she'd check the sock. No bank—just a sock in her artificial leg.

When I was still in elementary school, my older brothers, young teenagers, went with a group of boys to apply to be pinsetters, a decent if slightly dangerous job, in a new bowling alley about 5 miles north of us—not in our community. My oldest brother and another big boy knocked on the door. A White gentleman opened it and demanded what they wanted. They asked for jobs, and he yelled, "I don't hire any N-----s. Get out of here." They turned and left. It rattled me to hear about it. I knew I had to learn how to navigate that treatment when I left Goulds.

In elementary school, I figured out how to be a good student like my brothers. I was selected to be one of five students to be on a 30-minute radio program in Miami called Youth Ask Business. I was assigned a question by my teacher, Mrs. Russell. I had to memorize that question because we were going to be on stage with Black business executives on a Black radio station.

My grandmother had said I could participate, but whoever took me had to bring me back to my door. Mrs. Russell assured her she would.

Afterwards, Mrs. Russell pulled up in front of the house. I didn't want Mrs. Russell to walk me to the door because I was ashamed of the house. The door didn't close all the way and the front porch had a couple of holes in the wall. I knew my grandmother was waiting up for me because I could see the light of the kerosene lamp. My mother was probably working. My grandmother came to the door.

"I'll just run on up, Mrs. Russell," I said.

"No, Wilbert, I will walk you to the door because that was your grandmother's requirement," she said.

On the porch, my grandma said, "Hi baby, how was it?"

Mrs. Russell said, "Mrs. Williams, Wilbert did a good job. He remembered his question and he didn't stumble." As Mrs. Russell turned and went away, she said good night, but I was embarrassed. It really motivated me more to help my mother get money so we could build a new house.

In the 1940s many Blacks worked as migrant farm workers. In 1949 my mother's brother, P.J. Bryant, was a migrant worker, who had 14 children, eight boys and six girls. He convinced my mother our entire family should go with him, on the season, to work in the fields up north. We worked in Maryland, Delaware and Pennsylvania and we made good money picking green beans and potatoes in the fields.

School was starting in September and my mother told her brother we had to leave to get back into our classes. My mother argued with her brother, saying she didn't care what he said, her boys were going back to school. She stressed the only way we could improve ourselves and make something of our lives was with education.

For our return to Florida, my mother bought tickets for the train. She only purchased tickets for two older brothers, Alvin and Willie, and herself. She said to me and Henry, "You keep your mouth closed. Don't say a word." We were still small, only 9 and 8 years old. And for little Bobby, everyone knew he was very young and wouldn't have to pay. She played this little trick on the railroad to save money. My mother prepared finger food for us, enough to bring us back from Maryland to South Miami.

My mother had been sending the money we earned back to her father. They had poured the foundation for our new home before we left. My mother wanted us to build this home from the ground up and not owe anything when we moved into this home.

We thought when we got back from up north, we'd see our house. The brick walls would be up and the window frames in. When we returned, it wasn't finished. The money had been lost through family mismanagement. The weeds grew up in the foundation just as we had left it. Of course we cried, but our mother told us not to. That setback made us work even harder to catch up.

We were two weeks late getting back and school had started already. We hustled and we caught up with our schoolwork in creative ways. When we had a unit on Pythagorean Theorem, I used the example of a telephone pole with its guy wire. I labeled the post "a," the distance between the base of the pole and the guy wire anchored in the ground as "b" and length of the guy wire from the top of the pole to the ground as "c," using the formula: C-squared is equal to A-squared plus B-squared. I was proud when my teacher asked me to go the blackboard to explain my example and to compute either variable.

With Alvin and Willie as teenagers, my mother announced her three rules—no drinking, smoking, or gambling. She said if she ever caught us drinking or smoking, she'd send us to the Reformatory School. My mother didn't drink or smoke and certainly didn't gamble. She warned us if she caught us shooting dice in the pool room, shooting pool for money, then we'd be in trouble.

One time I went to the pool hall to play pool for money. I played seven games against an older man I did not know. He won every game and started to leave with my money, 25 cents per game. My friends objected—he had to give me a chance to win some back. We played again and I lost again. After being fleeced by a real pool shark, I adhered to Claude's rules.

Further, she didn't want us to get in trouble with the law. We were not to fight anybody outside of the house. She knew we brothers would have little spats. For example, Alvin and Willie would promise us something if we did their chores. Alvin was tricking us. When he didn't deliver on his promise, we wouldd fight. Protecting Henry, I would say I started the fight and took the punishment. We were not bad kids, but as boys, we fought each other often.

Another of Claude's rules was she didn't want us going across Highway A1A, the other side of the tracks. My mother believed that side of the tracks was a bad environment. But the movie theater and the only grocery store in our small community were across A1A, even the Post Office was on Highway A1A. She demanded if we went over there, we had to go in groups.

Before I was in junior high, one day my mother asked my younger brothers to go sit in another room, leaving the three oldest, Alvin, Willie and myself.

She began with "I know you boys are growing, I don't want you all to get yourself in a fix."

I, the youngest of the three of us, was only 11 or 12; my brothers had started high school. She reached in her bag and pulled out these packets of condoms and gave us each one, saying, "Don't get any girl in trouble so take one of these and keep it in your wallet."

"Claude," I said, "what are those things?"

My oldest brother said, "Wilbert, shut up." I argued I didn't want it. I didn't know what it was. Alvin said again, "Wilbert, shut up." So I shut up.

My brothers wouldn't tell me what the condom was for. On the school grounds, I asked my friends who explained it to me.

Even if my brothers wouldn't explain sex to me, I realized I could learn other things from them, so I started absorbing what they were learning. For example, in seventh grade in the biology class under a strict teacher, Mr. Charles Smith, we had to learn all the major bones in the body. Every student had to stand at Mr. Smith's skeleton poster and name them. At home, Alvin and Willie would hold a book for each other and they would study the bones and the major muscles in the body and memorize them. I sat there, listening, soaking it all, preparing for when I was in seventh grade.

Finally in 1952 our dream of a better home came to be a reality. We'd bought all the cement blocks for the walls. Next we spent the money for the window frames and had a company install them. The wood for the roof and roofing materials followed. Then it was time to start working inside. They installed drywall and plumbing. I was thrilled—we would have electricity and running water.

My mother didn't want any of the old furniture from the old house. She wanted all new. We only brought with us personal belongings, clothing and a few other small things. Everything else was purchased brand new from a big furniture company called the Jefferson Store. By 1952 segregation had eased a bit, so she took a bus some 22 to 23 miles from Goulds to buy all the furniture with cash to outfit our new home. She didn't want any bills when we moved into the house. We didn't have a television. I didn't care; I wasn't big into TV; Henry loved to watch cartoons at a neighbor's house.

In our new home, we had three bedrooms, her bedroom on the far side of the house and two bedrooms at the back. She put twin beds for the four oldest boys and my brother Bobby slept on the couch in the living room. At the Jefferson Store she bought the twin beds, an electric stove, and a refrigerator. She picked out a nice dining room table for all of us to sit around. She still worked as a domestic, but we were proud—we had all contributed to our home.

Although my brothers and I worked hard, we did have fun of course. In junior high there were sock hops, dances on Friday nights. All the kids were invited to a school gymnasium to dance to recorded music. Henry and I practiced dancing a slow two step with each other or a pillow or a broom to be ready for the sock hop. He and I loved Fats Domino, the Great Pretenders, and Otis Redding.

At our sock hops, we'd listen to Motown bands like The Platters and The Shirelles. We danced with girls to Nat King Cole, Harry Belafonte, Little Richard, and even Jerry Lee Lewis and Elvis. The dance's location rotated through the different schools, Homestead to Goulds to Richmond Heights.

Mimicking my older brothers, I participated in school sports and several other activities. I ran for and was elected a member of the Student Council. I served on the Patrol Force, just like today's Safety Patrol. Like Alvin and Willie, I joined Boy Scouts.

Alvin, the oldest, and Willie became Eagle Scouts. I joined because I wanted the uniform, which was a big deal for me. My mother used that as an incentive. My mother said she'd buy me the uniform if I did all my chores and a bit more.

At Boy Scout meetings, I learned the pledge: "On my honor I will do my duty to God and my country." The oath taught me about growing up to be respectful of ourselves and be respectful of others. Boy Scouts developed my love of country.

Our scoutmaster, Mr. George Cooper, was tough, like my mother. He could be compassionate, but he didn't play around. The boys who acted up during Boy Scout meetings had to go through "the beltline." The boys lined up in parallel lines and slapped the offender with their belt end, not the buckle. All the boys took their belts off and slapped the misbehaving boy as he ran down the line.

I went through the beltline only one time. One hit me on the back of my thigh and it hurt. I begged my brothers, *Please don't tell Claude.* My mother would have worn me out if she found out. When Alvin and Willie, my anchors, went away to college, I lost interest in Boy Scouts.

In his high school years, Alvin played football and ran track. Alvin played as a backup quarterback as well as being a model student.

They were on the debate team, too. My brothers were determined to succeed and get to college with scholarships.

When Alvin was a little boy, he got his hand caught in bicycle spokes; it almost took his finger off. My mother took him to Dade County Hospital where they waited for five hours before a doctor or nurse treated him. After that, he announced he'd study to be a doctor and serve the Black community. When he got ready to go to college in 1955, he said "I'm going to study biology as my foundation for medical school." Willie announced he would do the same and become a dentist.

In 1955, as an honors student, the school selected me and another student to lead the graduating seniors into commencement. Awards were given to the graduating seniors, but the top award was the Harold E. Kendall Award, the man we later worked for in his lime groves. The principal announced, "Ladies and gentlemen, the valedictorian in the class of 1955 and the recipient of the Harold Kendall award is Alvin Bryant."

I almost stood up and yelled, *That's my big brother*, but everybody knew already. When Willie came along the next year, he didn't finish number one, he finished number four in his class, graduating with honors and a partial scholarship. With Alvin off to Florida A&M, Henry and I were closer than ever.

In March 1956, when I was 15 years old, I was home with Willie. We still didn't have a television. My younger brother Henry was watching cartoons at the house of the family down the street. He played and roughhoused with a boy of this family.

That spring afternoon my youngest brother Larry ran into the house screaming to Willie and me something had happened to Henry. We didn't believe him, just a little boy, at first. He finally convinced us to go check. I waited in the street, ready to see if I needed to get my grandfather and his car. Willie walked out of their house with Henry in his arms.

"What's wrong?" I yelled. "Should I run and have Eugene take him to the hospital?"

"Nope," Willie said, "He's dead."

The five-year-old brother of the boy got scared when Henry wrestled with his brother. The little boy took his father's rifle and shot Henry in the back. The shot pierced Henry's heart and killed him instantly. My brother was not quite 14.

The day this happened, my mother's job as a domestic required her to stay at work that evening. She hadn't planned to come home until the next day at the end of that day's work. We called the neighbors who went over and got our biological father. Together he went with them to bring my mother back home.

The neighbors only told her one of her children had been involved in an accident. They wanted to bring her home before telling her.

I was so afraid. I prayed, *Please God let them tell her this before she got home.* I knew my mother would scream like she was about to die. They did not tell my mother until she got home that my brother Henry had been shot and killed. I put a pillow over my ears, but I still heard her screaming.

The death really disrupted our entire community; during those days of segregation, no White policeman came down to investigate. They must have decided an accidental death wasn't their problem.

We didn't have a lot of money at the time. Rather than have a wake at a funeral home, we brought my brother home. They displayed him in the living room in the house. School closed the next day so that we could have the funeral.

We came together as a family. My oldest brother Alvin was away completing his freshman year at Florida A&M University. He came home to take over and help my mother get my brother buried. The entire community gathered around us to try to lift us. Then Alvin went back to college.

Years later, I still mourned Henry by searching for his grave. The cemetery in Goulds had been closed because it was traditional, all across America, not to take care of Black cemeteries. The area gets run down and officials condemn them and close them. After the officials closed the cemetery I lost track of what happened to it. For years I looked to find where my brother was buried, but I never did.

I started high school that fall and I had to get ahold of my sadness, to find a way out of Goulds, and into my future.

Chapter Two

Coaches and Mentors

The summer before I entered high school, Alvin, Willie, Bobby, my uncle Eugene and I started to work for Mr. Herbert Graff, a Jewish flower grower near Goulds Canal. South Florida was an ideal place to grow flowers, and he grew asters, marigolds, snapdragons, lilies, daisies, and candy tufts. We planted, cultivated, weeded, and fertilized them.

Many times, Mr. Graff would drive to the edge of the school's baseball field, pick us up, and take us straight to his field. We worked weekdays from about 4 until 6:30 every evening, earning 50 cents an hour, another source of income for my family, especially when we weren't picking potatoes, tomatoes, and beans, or cutting pole bean sticks.

Weekend days, Mr. Graff would come to get us when we thought we were not expected. He'd blow his horn—beep—and tell us how many of us he needed. If we hadn't had any lunch, it didn't bother us. At 12 o' clock, Mr. Graff brought us into his home. We'd wash up. Mrs. Graff was the nicest lady you'd ever want to meet. I didn't see a color when I looked at Mr. and Mrs. Graff. They were like an uncle and an aunt to me, to us. Most of the time they'd have a big pot of soup with tomatoes, okra, and some kind of meat. They had cornbread and usually some fruit. We'd eat lunch around the table with them, Mr. Graff on one end and Mrs. Graff on the other end. At 12:30, he'd say, "Okay, boys, it's time to go back to work." His favorite saying was "Lawd-a-mercy, boys, I don't know if I'm gonna make it another day. But by God, we're gonna give it a try."

There were times when Mr. Graff didn't have funds to pay for our work because he had to sell the flowers first in the Miami market. I remember my mother asked him for some money for the boys' work these past two weeks. We trusted him; he was a man of his word.

He had a son, Herbie, who was in the Navy when we started working for him. He married a German girl and brought her home. For Mr. Graff that was a big faux pas, a mistake. I was just a boy so it was hard for me to understand why Mr. Graff was mad at him. He put Herbie Junior, his wife, and grandson in a one-room shack out in the middle of one of his fields: He would not let them live in the big house.

I had a talk with Mr. Graff one day about it. He told me Herbie let him down. I argued that Herbie was so nice. Mr. Graff explained it this way: he married a German girl. He asked if I knew what the Germans did? I had read about the Holocaust. Mr. Graff told me there was nothing wrong with liking Herbie, and he would always be his son.

That was the first time I'd ever seen anything like that. Even though we were Black and everything was segregated in those days, we had never experienced anything like that.

We boys were very competitive in sports. I wanted to be better than my brothers. I first became the trainer for the football team in junior high. I tried out for football in ninth grade. I wanted to play football because Alvin had been quarterback. During those days the quarterbacks called the plays. The smartest kid played at quarterback.

I wanted to be quarterback, but I didn't make the team. I went to the coach who I knew very well because I had been his trainer. I told the coach, I knew I wasn't as good Emory Collier, who was his starter, but I was the second best.

"Son," the coach said, "I want you to be my statistician. I want you to be my sports editor because you're a very good speaker. You're good with numbers. I want you to be prepared to get up and give the sports report before the entire student body every Friday during assembly." He told me I'd travel with the team, too.

Every Friday at assembly, I'd be announced: "Ladies and gentlemen, boys and girls, the sports report from Wilbert Bryant." I'd give the sports report to the entire student body. I had it all written down in advance, but I would speak extemporaneously. So that was my start in public speaking. I even wrote a column in the student newspaper on the sports report and it became very popular.

I thought if I couldn't play football, I'll go out for basketball. I tried out for the junior varsity basketball. On the junior varsity team, I was starting point guard and I was one of the stars. So here I am, my 10th grade year, and I'm working with JVs on one end and the varsity squad is on the other. I overheard the coach tell the varsity players, pointing to me, saying if they didn't play better, Wilbert Bryant would take one of their positions next year. When I heard that, I gave the game everything I had.

My second year playing junior varsity, Mays High School completed a brand-new gymnasium, which was a big deal in the

South during those days for a Black school to have a brand-new gymnasium floor with bright hardwood. We had played all our games on an asphalt outdoor court. For our outdoor games, they closed the school and put chairs all around the asphalt.

When eleventh grade rolled around, the coach named the starting five. I gained a spot as the starting point guard. I played my entire junior year as a starter on the basketball team.

In 1956, when I went to high school, my oldest brothers were away at college which left me in charge as the oldest in the house. I had to devise other ways to help my mother. I got a job in the cafeteria washing dishes Monday through Friday during our 30-minute lunch break. The first 20 minutes, I'd wash dishes. For the last 10 minutes I would eat and go to class. I had friends who laughed at me because I worked in the cafeteria. It didn't bother me one iota.

My job meant my mother didn't have to give me any money for lunch because money was tight. I kept this free lunch from my mother. I didn't tell her I washed dishes at school and she didn't have to pay for my lunch which was 25 cents by that time.

I had another job as well, washing dishes at La Casita Tea Room in Coconut Grove. My aunt Jimmie Lee Williams was the head cook and she helped me get this job. I washed dishes from nine in the morning until nine at night on Saturday and Sunday. Mr. Bob Osley, the original proprietor, instructed me how he wanted the dishes handled. He didn't want any dishes on the top shelf where the busboy placed the dirty dishes from the dining room. He didn't want any water spots on the glasses or the silverware. I was to use gloves to take the plates and glasses out.

Every weekend day, I earned $9 a day, which was a lot of money. My mother only made $6 a day at that time. On Sunday night, Mr.

Bob told me I had done a very good job. He paid me my $18, but he was so pleased with me he gave me a $5 tip. Five Dollar Tip! He was a Jewish proprietor, a wonderful boss just like Mr. Herbert Graff.

The kitchen would let me take biscuits, rolls or any leftovers as much as I could take. I would take at least two containers of food home to help my mother reduce her grocery bill.

Out on Highway 1, I'd stand in front of Poe Hardware store in a well-lit spot. I would stick my thumb out and I thumbed or hitchhiked home every Saturday and Sunday night. I never caught the bus. White or Black, people would pick me up. They'd ask, "where you going, boy," just like that. I'd say I was going to Goulds. They'd say they were going to Coral Gables or Homestead, that's south of Goulds. They'd give me a lift as far as they could on their route.

When I got home I was so happy as I told my mother what Mr. Bob gave me. She thought that was outstanding. My mother took my $18 and but told me to keep the $5 tip for my expenses during the week.

In senior year I was proud as can be to be starting point guard again. My classmates, the little girls and all their friends, came around because I was popular as a sports star. The girls wanted to give me an apple after the game and lots of attention. I had a girlfriend and she'd walk with me to practice. As a starter I knew I was in tall cotton. They weren't going to sit me on the bench.

Practice started at 3:30 sharp. The first time I got there at 3:37, the coach called me over and reminded me practice started at 3:30. He said, "Son, I don't tolerate that on my team. I want you to give me 15 suicides." The player had to drop into a three-point stance point and run from one pole to next and back 15 times before joining practice. That was my punishment.

The next week this little girl and I walked up about four minutes late. The coach gave me 25 suicides that time. He warned me not to be late. He said he wasn't going to tell me again. I was the starter, a star. I was the one dishing the ball around. I was sure he wasn't going to do anything to me.

The third time which was on a Wednesday and we had a game Friday night, I was about two minutes late. Practice had already started. They were doing calisthenics. Coach never said a word to me.

At Friday's game, they began to announce the visiting team. "Now your starting five from your Mays Rams." I jumped up and started taking off my warm-ups.

Coach looked down there and said, "Bryant, what you doin'?"

I said, "Sir, he's getting ready to call my name."

Coach said, "Sit your rear right down there. You're not starting."

I thought, *Oh my goodness.* The coach benched me for half of my senior year season. I came off the bench but I didn't start.

The last game of the year we had to play a team in West Palm Beach called Roosevelt High School, a big-time high school with a big coliseum that made our gym look like a cow pasture. You don't go to West Palm Beach and beat them on their own court.

Before the game started, the coach came into the locker room and told us he was going to start all the seniors in the last game. I was a senior getting ready to go away to college. Coach had a few words for each one of the seniors. I was the last senior he mentioned.

"Bryant," he said, "it seems you are a bright young man. I'm going to start you at point guard because you're a senior and this is your last high school game. Son, you had a very promising future. I probably could've gotten you a partial scholarship to a college if you had played up to your potential and not let those little girls get in your head and let you think that you're bigger than the team. I'm not gonna say very much but you have to abide by the rules and regulations."

Fast forward. The score was tied 66 to 66. We had about 40 seconds left on the clock. We were bringing the ball down the court. One of my guys took the shot on the other side of the court. West Palm Beach Roosevelt's best player got the rebound. He was a lefty. He made a mistake of dribbling with his right hand on the right side of the court. He was going to make a layup on the right side of the basket. He was 6 foot 3 inches, I was 5 feet 9.5 inches. We were running. I went under him, and I submarined him—I hit him with my shoulder. He fell on top of me. I lay on the court and played like I was knocked out cold. An offensive foul was called on him, but I had really fouled him. They came out with smelling salts. I came to. I was acting like I'd win an Oscar.

Their fans had been going crazy. They threw cans and wrappers and popcorn on the floor. The school had to bring the police to stop them.

The coach called timeout. I got up, and he said, "Son, are you okay?" I told him I was okay. He said, "Can you shoot those free throws or do I have to substitute someone?" I told him I was good.

The coach sent me to the free-throw line. 66 to 66. He reminded us there were eight seconds left on the clock. If I made the first free-throw, the score would be 67 to 66. I got up, pushed a throw and I made it: 67 Mays, 66 Roosevelt.

The other team called a timeout to ice me. Coach called us into the huddle. He said, "Bryant, you keep your concentration. We need you to make that basket. In the event they get the ball in that eight seconds, they can get the ball down the court and make a basket and tie up the game and at least we will go into overtime." The pressure didn't bother me because I can shoot free throws.

At the free throw line, the buzzer sounded. Whoosh! I made the second one. Roosevelt couldn't get the ball up the court before the eight seconds ran out. We won the game 68 to 66. *My goodness!* Coach told us to get in the locker room.

Usually after we had an away game, they would feed us and put us on the bus to South Miami. This time Coach hustled us to grab our clothes and warned us, "Don't even try to put them on." We had to lie down on the floor on the bus because they were throwing rocks. We took off heading back to south Miami. About 20 miles outside West Palm Beach Florida, the bus pulled up to a store like a 7-Eleven and we went and bought some Nab cookies and some sodas. But we had won.

The coach taught me that I'm not bigger than the team. I'm a part of the team. That major lesson in life has served me to this day. No matter how good you are, you have to abide by the rules of whatever organization you're in. No one is irreplaceable. Coach James Anders, a former varsity letterman in basketball and football at Florida A&M: I loved that man.

Our social life in the 1940s and '50s was limited, but we found ways to amuse ourselves. I had two very good friends, James Perry and Curtis Armstrong. None of us smoked or drank. Their parents had many of the same rules as my mother, although one of them would sneak in a smoke now and then. I knew that my mother would detect the smell so I never tried. I knew my father, who lived on the next street, smoked cigars and I always wondered what that was like.

I went to the dances when I could. I planned in advance for our senior prom, our big memorable event. A girl, Shirley Howard, moved into Goulds in seventh grade and we thought she was the most beautiful girl in Florida. Every boy at the school had a crush on her, including myself. I knew I didn't stand a chance because first, I didn't have an automobile. Secondly I didn't have any money to spend on a girl to take her out for a soda or cookies. I never approached her until February of my senior year.

Shirley was coming down the hall when I asked her if she was going to the senior prom this year. She said she'd like to go. When I asked her if she had a date, she said no.

"Would you like to go to the prom with me?" I asked. She said she'd be honored to go to the prom with me. I thought this was fantastic. I next asked her for a favor: Don't let anybody know that I was taking her to prom. She was surprised, but she agreed. All the other big-time athletes competed for her.

When the time approached for my senior prom, I went over to my father's home to visit him, even though he had disowned us. I told him the prom was coming up and I wanted to use his brand-new blue Ford to drive my date to the prom. He asked, if he let me use it, what time did I anticipate returning it. I explained the prom ended at 11 o'clock. All of us would go to the Virginia Key Beach, the only beach Blacks were allowed to visit and have fun. It had a jukebox on a little cement court for us to put a coin in and listen to music. I told him the car would be home by 3 a.m.

My father said, "Okay, but don't wreck my car." I said, "No, sir." He said, "No drinking."

"No sir, I don't drink."

On prom night my best friend James Perry and I went to pick up our dates. Shirley was my date and Jim's date was Helen Davis. At the prom when we walked in the main entrance, it seemed like all the music stopped and it was like they turned the lights up a little more. Everybody looked and saw me coming in with this beautiful girl.

When Shirley and Helen went over to get some punch and cookies, Harry Jones, one of the tackles on the football team, came over and grabbed me by my collar, saying "Bryant, are you trying to get Shirley for yourself? I'll knock you out for bringing her in here. I gonna beat you up next week."

"Harry, turn me loose," I said, "You're messing up my shirt." I had on a nice white shirt and a tuxedo. "You try that. You know I have a lot of brothers and they're gonna come and get you." This guy had gone crazy.

Anyway, Shirley and I danced two or three dances. When I went to the other side of the room, another big guy, one of the big guards on the football team, grabbed me the same way. He liked her, too. He threatened to beat me up after graduation. Two guys promised to beat me. *Oh, my goodness.* My two big brothers had graduated and were away at college, but my uncle was there. I figured the guys were faking. I didn't really think they were gonna punch me, but it added some drama to our prom.

My girlfriend was on the sidelines when I walked in with Shirley, and she said some bad things that I ignored. I had given this girl my class ring and asked her to wait for me until I finished college. I had told her I was taking Shirley to prom as a friend. That was the truth.

Anyway, we came back and took the car to my father. I thanked him. Loaning me his car was the only thing that he did for me in all my life up to that point.

In spring and summer of my junior and senior years I played right field on the baseball team. I wasn't good enough to be a starter, but I came off the bench and played in every game. We had a great group of players.

Our last home game before the state tournament was a night game against Fort Pierce, Lincoln Park Academy. We played in Homestead in a White-owned park that let Black kids play at night. The winner of this game would get a bye in the first round of the tournament.

In that game in the last inning, we were down one run. We had runners on first and second with two outs. Coach Earl Dinkins wanted me to pinch hit for the next batter to hit a single with my strength, so our runner on second could come around and score and tie the game up.

The coach told me that this pitcher, a south paw, a lefty, had two pitches, a slow curveball and a moderate fastball. He always threw the fastball first, then a slow curve on the second. Coach warned me not to try to hit it out of the park.

The ball coming out under the lights looked like a grapefruit. Whish—strike one. I missed it. Oh my god! Next was the fastball—I was sure I'd get it. I didn't.

The coach jumped up and threw his hat and said some very colorful words, yelling at me, "Don't kill the ball!" Coach warned me the pitcher was coming back with this slow curve this time. It was zero and two on me, no balls, two strikes. The slow curve would be next. The coach asked, "Do you understand? Just put it in play."

I answered, "Yes, sir." I moved up and backed out of the box to get the pitcher a little rattled out there. Back up to the plate, I'm in my stance. I see the ball coming, another slow curve, another big grapefruit. I knew I had it this time. Wham, another strike, three. Coach used some more colorful language. He threatened to never put me in to pinch hit again. I told him I was sorry. I just knew I had it. I had known I could hit at least a double or a triple, but the coach was right. I didn't.

Coach said, "I didn't want you to kill it!"

For the last game of the state tournament, we went to Bethune Cookman College where the games were being played. In the final game, we ended up playing a big high school out of Jacksonville, New Stanton High School. Baseball great Hank Aaron's brother-in-law played shortstop on that team. They came out looking good in brand-new blue-and-white uniforms. Our colors were scarlet and gray, compared to them, we looked like a ragtag bunch of guys, but we were very good players. We went to that final game and we beat them, 12 to 3. I played in every game of the tournament. My junior year we won the state baseball championship in Florida for all the Black high schools, my school's only state championship. What carried us to the championship was our two great pitchers, L. Jo Bohler and Dennis Bohler, two blood brothers, a lefty and a righty.

Coach Earl Dinkins, who had been a great player for Morris Brown College in Atlanta, was our head coach. Could he use colorful language! Another lesson learned: Listen to the coach.

When the Brown versus the Board of Education decision was rendered in May 1954, it had no impact whatsoever on our school. I graduated in 1958, No. 11 in my class, good but not enough for scholarship money, which stopped at No. 5. I told my mother I wanted to attend college like my two brothers. She asked if I knew where and I told her Florida A&M University in Tallahassee to be with my big brothers. Claude agreed I had done well in school.

I was going to have to work and pay for my way for my four years in college. My mother helped out. She knew the vice president of a bank in Homestead, Mrs. Mamie P. Smith. My mother had worked for her. My mother told her we had saved, but we didn't have enough. Mrs. Smith gave my mother a loan of $1,200, a lot of money in those days, enough money to pay for my first year and help Willie with expenses. Claude believed being in debt for education was all right but not for a house.

My mother told me I had to get a job at Florida A&M because she couldn't pay for my education after my freshman year. I told Claude to get me there and I will find a way to stay a second year. That's the deal we worked out.

Chapter Three

Straight Outta Goulds

When I was getting ready to go to Florida A&M, my mother asked what I was going to study. I told her mathematics. My mother argued all I could do with a math degree was teach school which doesn't earn much money. She wanted me to become a doctor or a dentist like my brothers. I didn't want to be a doctor or dentist and repeat what my brothers were doing. I told her I had read books and magazines in the library where mathematicians were needed in the corporate world like RCA. My mother didn't know anything about that. I insisted I was not going to teach school. I would get the bachelor of science degree that required certification and have it as a fallback option. She told me if I didn't finish in four years, I would have to come home to work the fields. She reminded me of my other brothers and sister coming along. I told her not to worry about me. I planned to graduate on time.

Why math? I didn't want to be around the sight of blood, and I had a crush on my freshman year high school algebra teacher. Her policy was the student who got the highest score on the weekly test got to lead the class to lunch. She would put her arm around the shoulder of that lucky kid. She smelled like flowers, like lilacs.

I only applied to Florida A&M University in Tallahassee. One reason was sports; Florida A&M came down to Miami and played in a football game called the Orange Blossom Classic. At that time, it was the biggest Black college football game held in America. We had 37,000 to 40,000 fans in the Orange Bowl. The Great Marching 100 Band performed. In the 1950s and '60s Florida A&M was one of the best Historically Black Colleges and Universities in the country.

To be at Florida A&M was thrilling and intimidating. Seeing those big buildings and all these kids coming from all over the state and all over the U.S. startled me. But I thought if my brothers were doing okay, I'll be okay, because I'm just as smart as they are.

Florida A&M, as a land grant university, required all young men to be in ROTC, Reserve Officer Training Corps, for two years and enroll in Military Science 1 and 2. I couldn't believe both my brothers were about to become second lieutenants, which was a big deal for a country boy coming from down South.

Two family incidents from my very young childhood guided me on this path as well. My cousin Thomas Williams came home from the war. He was in the Navy and wore this sharp white uniform. I think my uncle was over 6 feet tall. I, a little guy, was so impressed with him. Another time my mother's sister and her brother-in-law came to visit us. They had a son who was in the Army. My cousin sent back a picture of him in his army uniform. He teased us, but he really encouraged us to get a good education.

As soon as I showed up at Florida A&M, I was looking for a job. My second day there, my big brother took me over to the director of student activities, the Rev. M.G. Miles, who lined up all the jobs on campus. Alvin introduced me and said I needed a job.

The director laughed, saying, "another one of those Bryant boys." He asked me if I could type. I told him yes. In high school, I had taken typing as an elective, even though boys usually didn't. I didn't get the A I earned because the teacher thought I'd never use typing. Boy, she was wrong. The Rev. Miles needed me to type 35 correct words per minute. I knew I could do much better than that. His secretary gave me a typing test. I typed 48 correct words per minute. She was impressed and told the director. He sent me to the office to Mr. E. M. Thorpe, the registrar of the university, who would employ me. That was my first job, 10 hours a week. I set my own schedule based on my academic load.

I knew how hard my mother had worked to get me there, and I didn't want to let her down. Plus, majoring in mathematics and getting my minor in education required a lot of study. Most semesters I took only 17 credit hours because I was in college algebra, Euclidean geometry, and math courses like that.

Many students have a rude awakening early in college, showing them how different it is from high school. I turned in my first paper in English composition, certain I had done a great job. Imagine my shock when I opened it—a big F and "see me" written on it in red ink.

My professor said to me, "Do you know Mr. Alvin Bryant?" I answered he was my oldest brother. She said, "I'm ashamed of you. Go see your brother about this."

At 10:30 that night, Alvin sat me down and said even he was ashamed of me. He demanded I rewrite the paper right then. I had to do three drafts and didn't finish until 2 a.m. I certainly got the message about college studying right away.

Another reason I studied was because I knew at the end of freshman year, all math majors records were reviewed. If the math department didn't think a student had the aptitude for mathematics, he or she were pushed out. I didn't want that to happen to me. Bryant boys didn't fail, so I really hit the books.

There were many times my freshman year when my friends came by. They would say they were going to the movies over in Tallahassee. I suggested we go on the weekend. They teased me that I just want to study because I wanted to pledge Alpha like my brothers.

They were right: I wanted to be invited into Alpha Phi Alpha, the top fraternity at Florida A&M with over 50 members. My oldest brother was a senior, my second brother a junior, and both were members. APA emphasized scholarship and high academic standards and gentlemanly behavior, including wearing a coat and tie to class. Most of the members assumed leadership positions on the campus.

Every day in math class there was a spot quiz, and I didn't like to miss any problems. I had to practice and work problems and learn theorems. I told my pals you can't fake in math, an exact science. So they left me alone. From Monday to until around 6 o'clock on Friday I was all about studying.

I prepared to take my first exams. Since I usually did very well on exams, I didn't worry. I was enrolled in an elective entitled Elements of Biology, a three-hour course. I thought that I was well-prepared because of what I had been taught in junior and senior high school. In this class, we studied the plant kingdom. All it took was some serious study to memorize the material.

Professor Brown gave our class our first exam. We had 125 students in that classroom, big even for the university; I'd never been in a class that large. The exam was on Friday. When I walked out, I felt confident. I had done well.

The following Monday, the professor warned us first thing that some of us were not going to make it out of this class. The great majority had earned a D or an F on this first exam. On the exam only eight students earned a passing grade. When he announced that, I almost stopped breathing. It had been a humdinger of an exam, but I thought it was a piece of cake because I had studied. I had taken notes and rewritten them and memorized the material. I would go to the men's room about 6 o'clock in the evening when most of the boys would be out of the dormitory and recite everything I had memorized.

Dr. Brown cleared the first row and called down the eight, one by one to take a seat. He wanted the serious students to be in the first row. Dr. Brown called five names. My name was not among those first five. On the sixth name—I will never forget—he called Mr. Wilbert Bryant. He gave me my papers. My heart was pounding, number one because I was worried that I was not among that eight. I was sure I had aced that exam. I opened it up—an 86—I had a B. After class, a pretty girl, Yvette, tried to get me to study with her, but I told her no, I study alone. Anyway, passing that exam was a big experience. I studied hard. I had a good memory. As a result, I earned a B+ in that three-hour class.

My freshman year, a bunch of my home boys would all come together, and we'd socialize. We met an equal number of girls. The guys asked me to make friends with a girl. I had a girlfriend back home and I did not intend to break up. Still, I'd go to the movies with this college girl. After three weeks I knew she was not my type.

She was eager for more intimacy or what we called fresh. My mother told me she would have none of that. I had to get my education.

On Friday nights we had the movies, and on Saturdays we had a football game. Or if the team was away, there was always ping-pong in the student union. We'd go out on the intramural field and play football or softball. I still didn't smoke, drink, or gamble. I didn't buy any of the Boone's Farm wine the guys were drinking. On Sundays, I always went to church.

I had several friends: John Brown, my cousin Bennie Gibbons, Batey Woodward. One special friend was Helen Davis from my high school. Helen knew of my money challenges. In those first two years, every time she received a package from her mother with food, chicken, cake, or money, she would call me to come get some before her dormmates would ask for it. I'd run over and run back to my dorm. If her mother sent her five dollars, she would give me two of the five. We became best friends. I carried her messages to the starting quarterback who became my fraternity brother. I took her to the ROTC ball sophomore year, one of the biggest balls on the campus. I wore my Army uniform, and she wore an evening dress. To this day, she is one of my best friends.

In student activities. I was a member of the men's senate and was involved in the NAACP. I ran for office, and they selected me to be treasurer.

I had done well my first semester. The Alpha Phi Alpha fraternity recognized the fact that I had a B average at that point, plus I had the Bryant name reputation. I was invited to their Smoker for the young men to get to know us for possible membership. My second brother would be there during the year that I would be pledging in the fraternity. So they extended an invitation for me to pledge Alpha

Phi Alpha. You had to have at least a 2.8 GPA grade point average to pledge. In the fraternity, your grade point average could not drop below 2.7, which was a strong C+ average. I was determined to keep my grades up. I made the fraternity in April 1960, a big accomplishment for me.

My sophomore year rolled around with the same schedule—study, games, dances. I was on the ROTC Drill Team, where I learned how to handle my Army weapon, a skill that served me well in my Army career.

At the end of my sophomore year, the Advanced Course Exam was administered to all the young men. The Department of Military Science used it to look for good Army officers. Every young man was required to take four semesters of military science or face the draft. If you didn't get in the advanced course, you would be drafted.

I earned the eighth best score in my class, so I was selected to go into the advanced course. I was headed toward becoming a second lieutenant in the United States Army like my two older brothers. This meant I would have a leadership position my junior year. When I came back for my junior year, I wore insignia on my left collar that indicated that I was in the advanced course.

In my sophomore year, I continued working in the registrar's office as well as collecting tickets at the movies on Friday nights as my second job, my second paycheck. At the start of my junior year, I picked up two more checks. For Advanced ROTC I received a $30 check. After they took taxes out, I got $27.90 a month.

The dean of men asked me to be an RA, a resident assistant. I would have 40 young men on my dorm wing. It was another big check coming in, my fourth paycheck. I couldn't believe it because both my older brothers had held similar positions. Here I was following in their footsteps.

Even though I was not required to have a roommate, I asked if I could have a roommate who might be having difficulty in math. The dean had a student like that, a Nigerian student named John Isayu Makuku.

In my first encounter giving back to education, I provided John basic instruction in math for him to earn a C or better. We agreed we'd be back in the room by 6 o'clock so we could study between six and eight. I worked on his basic math, for example linear equations with one unknown like $2X + 7 = 5$. I had a little chalkboard. I told him to memorize facts. I had him rework problems and I'd check them. I felt good at the end of that semester because he passed both classes, math and English.

When I got the fourth check, I called my mother with my good news. I told her she didn't have to send me any more money. With my summer earnings, my four jobs on campus, I could pay my own way. In fact, I only needed two checks to pay for my expenses here on campus; the other two checks put money in my pocket. My mother told me off: I didn't tell her what to do. She'd send money if she wanted to. I agreed but I said, "Don't send me a lot." My mother would sometimes send me five or six wrinkled one dollar bills in her letters. That's mothers. The conversation taught me a lesson in how to treat my children if I had them, or nieces and nephews who were trying to get an education. I would try to emulate my mother and treat them well. Give them incentives to try to work hard.

At the end of junior year in order to be commissioned as second lieutenants at graduation, a six-week ROTC summer camp was required. At ROTC camp at Fort Benning, Georgia, I learned all the basics, small unit tactics, communication, map reading, everything military.

I had never been out of the state of Florida. I called my oldest brother and my second brother. They had given me good advice on how to make it in college and now they helped me prepare for camp. The key was to study. Don't play around and drink beer with the other guys down at the canteen. They warned me I would get a little blue slip in my box one day telling me that I would be a leader of a company of 220 men the next day, and I had to be ready to perform.

Early in Summer Camp, I was put in a leadership position, the company commander in charge of all those 220 men. It was very difficult for a young Black from the South to lead a company of 220 men, where 218 of the men were Caucasian, only two were Black. I was in the first platoon and my classmate from Florida A&M, Jackie Caynon, was in the second platoon. I saw him on the first day and didn't see him again until we graduated from Summer Camp.

My task as commander was to move them from point A to point B to point C and on, all throughout the day. With my self confidence and my experience of the ROTC drill team, I knew how to give commands in a command voice.

For entertainment, we had movies at Theater No. 11 in Harmony Church at Fort Benning. After my training week, 10 or 11 guys asked me to go to the movies with them. I told them I had to study. They insisted. To make a long story short, they would not go unless I went with them. There were about 11 of us and I was the only Black in that group. They'd put me up front because they thought I knew everything. When we'd go, we had a great time.

When classes started, I found the color of the other guys didn't bother me because I had always been taught by my mother that God created all of us in different shades. Mr. Graff and Mr. Osley treated me with respect, paving the way for me at Fort Benning. My mother always said to respect yourself, respect other people. Most of the time you get that respect back. That's the way I operated. I don't lie, cheat or steal. Thanks to Mr. Graff and Mr. Osley and my mother, I learned to only see Army Green.

At the end of the six weeks, I finished in the top 20 in the company. There I was, this young Black guy, coming out of south Florida and Florida A&M. We had kids from the Citadel, North Georgia Military College, Ole Miss, Georgia Tech and from all the southern schools. What a great experience that was.

Even though the military had desegregated under President Truman via Executive Order 9951 in 1948, integration hadn't really hit the Army the way he intended it to. I had no problems because my uniforms were always prepared. My shoes were shined. I didn't get any demerits or do anything to get written up. Because I knew what I was doing, the guys always wanted me to be a part of the group. We had so much fun for six weeks.

At the end of camp, we Florida A&M students went home for the remainder of the summer. All of us had done very well. We thought we had one of the greatest ROTC classes to come out of Florida A&M. When we returned to the university, I wondered what position I was going to have for my senior year at Florida A&M, based on my outstanding performance in that 220-man Summer Camp company. The assignments were posted in the ROTC building. There it was: Wilbert Bryant, company commander, Company A. I didn't know at that time, but to be company commander in charge of 125 cadets was the ideal job for me.

Chapter Four

Romancing Emily

As my college career was coming to a close and my military career was about to launch, my true life was beginning. I met Emily.

In summer 1962, about two months before I was to graduate, I looked out my dorm's window with my friends and here came a group of five girls on the sidewalk. I asked my friends who was the pretty, deep brown-skinned girl with the long black hair on the end. One guy told me it was Emily Mitchell. They'd been in some classes together in Science Hall. I wondered if she was a nice girl. Did she have a boyfriend? The guy chuckled, saying she didn't fool around.

I sent messages to Emily for two weeks. The summer session was only six weeks long. I didn't have a lot of time. One day I was walking up the hill to the dining room, and here came Emily. She recognized me. I didn't realize she knew me.

"Hey you, just a minute," Emily said. "Is your name Wilbert Bryant?" I nodded. She said, "You're the boy that's always sending me these messages. What do you want? Why can't you just deliver your own messages?"

"I'd like to get to know you," I said. "I'd like to come see you. Maybe go to the movies." She wasn't sure about that. I told her I'd like to call her in a couple of days.

"You can call me. I may not be there," Emily said. I thought, *Oh my, this might really be the girl for me.*

In college I already had my heart broken a couple of times, so I had given up on girls until I met her. In high school I had given my girlfriend my class ring and asked her to wait for me. She didn't. Somehow I knew this Emily was different. Sure, she was pretty, neat, but very focused. She knew what she wanted to do. I soon found out she was totally different from the other girls that I had met at Florida A&M and in high school. A city girl from Savannah, Georgia, she had patience to listen to what I, a country boy, had to say. I was a senior and maybe she knew I was going to be commissioned a second lieutenant. Maybe I was a good catch. I didn't know.

I soon found out she was very ambitious and career minded. Her maturity came from her grandmother who guided her for the first eight years of her life. Her grandmother instilled her values and teachings. She was wise and gave Emily chores and responsibilities that most five or six year olds never get.

A day or so later, I called and asked if I could come see her. She came outside and we sat on a bench in front of the dormitory. Emily said, "So what do you want to talk about?"

"I've seen you on the campus and I wanted to introduce myself. I'd like to see you and have some fun." She said, "Okay."

We weren't locked in exclusively as a couple at that time. During those days, girls graduating from college were trying to find husbands. A nursing student, Ima Wiggins, supposedly had been watching me and asked me to take her over to a party at the senior dormitory, Truth Hall. She was a nice girl, but I really didn't know her. When we walked in Truth Hall who was the first person I see: Emily! She was there with four other girls. I thought, *Oh my god, I cannot allow Emily to see me over here with this girl.* I told Ima I had to go to the men's room. I tried to think what to do. I decided to fake a headache.

When Ima went over to get some punch and cookies, I spoke to Emily. As we danced, she said, "Oh, I didn't know you were going to be here."

I whispered I wasn't going to be here very long. Emily was such a social person, chatting up all her friends, not paying much attention to me. This time it made me happy she didn't. I went to Ima and told her I had to bow out and go back to the dormitory and get myself an aspirin and lie down. Ima said she was sorry and asked me to give her a call tomorrow. I wasn't going to call or see her anymore, but I wanted to be respectful. I called her the next day and told her I couldn't see her anymore. I really didn't want to jeopardize my chances with Emily. That's how Emily and I got started.

In August 1962, two people were at my graduation: Alvin and Emily. She'd stayed to attend my graduation. My mother couldn't come because she was in the hospital.

When I received my degree, I went back to my seat and took my robe off. I had my brand new tropical worsted tan Army uniform on under my robe. I reached down for my service cap because it was time for the commissioning ceremony for the ROTC cadets who had successfully completed their four years. I had received Infantry as my branch designation. Oh, I was a proud young man. I had spent about two hours shining my shoes and everything was almost in tip-top shape. We stood, about 11 of us. They called my name and I went up, and I saluted the colonel, who presented us with our commission.

A tradition for newly commissioned officers is when the first enlisted man salutes you, you have to give him a dollar. I had this dollar in my pocket. A sergeant came up and said, "Congratulations, Lieutenant," and saluted me. I saluted him and I gave him the dollar.

Emily has always loved to give gifts. She gave me the best gift, a man's jewelry box, black with a gold line around the top because my fraternity colors were black and gold. I thought, *Oh my, this is wonderful.* I knew I was falling in love with her. I thanked Emily, and she got on the train and went home to Savannah.

That left me with my brother, Alvin, who had driven down from Fort Dix, New Jersey. He was going to south Florida and I to New

Jersey, where my mother was living to be near her sister. I told my oldest brother I loved this girl and I thought I wanted to marry her.

That's when he said, "Wilbert, is this girl coming back to school here at Florida A&M?"

I answered, "Yes, she'll be back to finish her degree." She had stayed out one semester when a hurricane had happened during her junior or sophomore year and her family needed the money. She got a job and saved her money to return to college.

"Wilbert," Alvin said, "you might as well forget her. She is too pretty and somebody's going to take her from you if you aren't going to be here."

"She told me that she was going to wait for me," I said. "I plan to come back and see her."

Alvin, always blunt, said, "Forget about her and get on the bus." He took my footlocker and threw it in the storage compartment under the bus. "Get your ticket and forget about this girl."

I started crying. Here I am, a 22-year-old man crying because my brother told me that this girl was not going to be waiting for me. I said, "She's not like that. She's different from any girl I have met in my life."

He only said, "Get on the bus."

I must have cried half the way to New Jersey.

When I got to Trenton, my brother Bobby picked me up and we drove straight to the hospital to see my mother. In my mother's room was this huge banner reading, "Congratulations Wilbert." Oh, it really touched me. I hugged my mother and I gave her my degree. I said, "This is yours."

All the nurses and female attendants, a lot of young ladies, were standing on the other side of the bed. My mother turned to them and said, "Okay y'all can leave now. He's not looking for a wife." The girls had set up punch and cookies out in the lobby area for me. She cautioned me, "Don't you go talking to any of those girls."

That fall, I was communicating with Emily, but I had to patch up a little misunderstanding on a weekend flying visit to Tallahassee from Fort Benning in October, my last weekend before basic training started. Another girl I knew had been telling Emily stuff so she'd break up with me. I thought I really loved this girl, so I corrected all that at my first chance.

Later that fall, I called her and asked her if she could come up and visit me for Thanksgiving. My roommate Matthew had a girlfriend, too, so I thought we'd team up. Emily came to visit me. I put her up in the visiting officer quarters on post. We had some fun until late evening.

Emily announced at 11:30 it was time for bed and she'd see me tomorrow. I'd said I'd her prefer to stay with her. She sent me back to my room. She wasn't that kind of girl. She was a nice girl, not a fast one. I really liked that in her.

As we were getting close to Christmas. I wanted to meet Emily's mother and her stepfather, Daniel Singleton. I offered to pick her up in Tallahassee and take her home to Savannah. She told me when to pick her up. I had on my best uniform. I drove my brand-new SuperSport Impala Chevrolet to the dormitory where she lived.

At the reception desk I told the lady behind the counter I was here to pick up Emily Mitchell. At that time at HBCUs, girls had to be signed out when they left the dorm for any reason unless they were going to class. The lady told me Emily Mitchell signed out two days ago. I asked her to please check again and sure enough Emily had signed out two days ago. Wow—was I embarrassed.

What would I do now? I didn't have Emily's home phone number or address. I told the lady thank you very much and have a good Christmas. As I walked out of the dorm, I got an idea. At almost 5:00, I went straight to the registrar's office. The registrar, Mr. E.M. Thorpe, was the last person in the office.

"Wilbert, it's good to see you." He said. "First of all, congratulations," pointing to my uniform. "What brings you here?"

I had worked for him directly my last year as the top work-study student and ran errands for him and drove his wife and children home from school and did whatever he needed me to do. I said, "I need an address for Miss Emily Mitchell who lives in Savannah, Georgia."

"That's a problem. You know her information is confidential. Say," he said, "could you do me a favor? I want you to watch the office for me while I go to the men's room. Then I'll be heading home."

I said, "Certainly." I knew what he was doing when he stepped out. I went right to the files and found her address, 24E 32nd Street, Savannah, and phone number, 236-7572. When he returned, I said, "Thank you so much. It's good to see you. Please convey my regards to Mrs. Thorpe and the children. Give them all my best."

With a bunch of quarters, I found a phone booth and called Emily's home. A lady answered, I said, "This is Lieutenant Wilbert Bryant, may I speak to Emily?" Emily's mother answered and told me she was out with her friends. She asked if I was that lieutenant Emily had been talking about. I said yes and asked if I could visit that evening and perhaps stay until Sunday. She agreed.

I had never driven from Tallahassee to Savannah. This was the first car I had ever owned. On the main road was a young white sailor who had his thumb out, trying to hitch a ride. Men in uniform always helped each other out; the service or the color of their skin didn't make a difference. I did wonder why he was wearing a white uniform, but he was.

I pulled up and said, "Sailor, where are you going?"

"Savannah, Georgia."

I said, "Get in! Do you know Savannah?"

"I was born there. It's my hometown," he said with a smile.

When we got there, he proceeded to give me exact directions to Emily's neighborhood. I repeated them back to him and he gave me a salute. We wished each other a Merry Christmas. He hopped out as his folks lived nearby.

At Emily's home, I did something that was inappropriate. It was about 11 o'clock at night and I blew my horn.

A lady came out on the second floor porch and said, "Oh you must be the lieutenant?" I told her yes. She answered "You're at the right place. Come in. Emily is out with her friends."

Her mother put me across the house in a separate bedroom from all the others. I guess that was the room for boys who would come to visit their daughters. Emily was the oldest child in the family. I heard her come in late, but I didn't get up. My door was partially closed. The next morning I asked her why she left Tallahassee early when she knew I was coming to pick her up. Emily said she came home for a cousin's funeral. I knew she wasn't telling the truth. Later I realized why she had left before I arrived. She had friends she wanted to hang out with and my being there would interfere with her short vacation. Me along—that was like bringing a sandwich to a banquet. I was so focused on her, I didn't care.

Later that day a young man visited. Emily had three sisters, teenagers, at home and all they did was giggle. Everybody left me in the living room and they went in the dining room with this guy. I said to myself, "I'm not going to move." If this guy wanted to start something, we were going to fight. I'd never been in a fight over anybody, but I felt that strongly about her. It seemed like they were in there for an eternity. It was only about 15 to 20 minutes. When Emily returned, I asked, "Was this guy your boyfriend?" She told me he was a friend. She'd been president of the student council in high school and had been very active in school, so she had lots of friends.

I said, "If you don't want me to stay, I'll leave."

She said, "I did not ask you to leave." I said okay. I stayed there with them two to three days before I returned to Fort Benning. At the end of 1962, I knew I was very serious about her.

In December 1962, I took Emily to meet my mother because I wanted my mother's blessing. I thought everything went well. After the visit I called my mother, I asked "Claude, what do you think of Emily?"

"She's a nice girl but she seemed like she might be a little fresh. You really don't need to get married right now. You should probably wait two or three years."

"No," I said, "she's not like that. She's a nice girl."

My mother said, "She wears her dresses too tight."

I knew that they weren't. It was the current style. I told my mother she was a very nice girl who was Episcopalian. Very career oriented, Emily really wanted to make something of her life. I'd never met anyone like her before. Claude said okay, but she wasn't sure. I told her, "You have to believe me on this. I don't think I'll make a mistake of this." My mother still thought I should wait at least two years.

Around Florida A&M's spring break, a buddy of mine told me if I was serious about her, I'd better go get an engagement ring and propose to her. I asked him to go with me to Savannah because I had never done anything like this before. We drove to Savannah, picked up Emily and were riding around in the car. At 11:30 my buddy prompted me: "Didn't you have something to say to Emily?"

We were in the back seat when I proposed to her and told her I loved her and I wanted to marry her. I pulled the ring out.

I paid $199.95 for this engagement ring. The payment on it was $9.95 a month. I could have paid for it because I had the money saved from my TDY (temporary duty pay) and my per diem, so I did pay it off early.

She accepted but she posed one problem: I had to get her dad's permission, her biological father's blessing.

I visited Savannah to meet Emily's mother again to ask if I could marry her daughter. She gave me her blessing but reminded me I had to see her biological father who lived in Marathon, Florida. Emily was the oldest of six daughters, one with her mother and five daughters with his second wife. Her mother warned me he really loved Emily. He had protected her like a prison guard at Rikers Island Prison. I thought this was bad because this man was a brick mason. A big man physically, he was a partner in a construction team who had built most of the buildings in Marathon, Florida, the houses and the stores.

In my Chevy Super Sport Impala, I picked Emily up and we stopped in Miami at my oldest brother and his new wife's home.

Alvin said, "We're going to put you back in this room and put Emily all the way down the hall. Don't you be going in there." I told him he didn't have to worry about that—Emily wouldn't allow that.

The next morning we drove to Marathon. I'm the type who never got ruffled and I don't get nervous about hardly anything. Anyway, her father was a challenge because I had heard about this man with his big strong hands and his shotgun. I wanted to be on my best behavior when meeting her dad. I made sure my shirt was tucked in my trousers, neatly pressed, my shoes were shined, and my gig line was straight—buttons lined up to the belt buckle the way we do in the military. He had been in World War II as a private. This man would be checking me out. When we pulled up there in my new Chevy Super Sport Impala at his home, he was outside. He watched my every move. I felt like I was on stage. Emily introduced me, "Dad, this is my friend, Lt. Wilbert Bryant and he wanted to talk to you. Bryant, this is my father, John G. Williams."

He was certainly a big man, broad shouldered with hands so big he could probably hold two bricks at once.

"Sir, I want to talk to you about Emily. We've been dating each other for almost a year."

"Okay let's go down to the lounge and have a drink and you can tell me what you want to ask me." I told him I don't drink. He ignored that. "Let's go and have a drink," he said again.

I just said, "Yes, sir." Here I am, almost 23 years old and I had never had a drink in my life, sitting on one of the bar stools in this lounge. He ordered and poured me about two fingers of bourbon in my glass. He said, "Now, Lieutenant, what was it you want to talk about my daughter?"

I was nervous, but I don't usually get nervous for anybody. I composed myself and started in. "Sir, Emily and I have known each other since August last year. I love your daughter and I'm here to ask for your blessing to marry her. Before you answer, I promise you I will never abuse your daughter. I would never put my hands on her. I would never fight her. I would try to give her the most respect at all times. I will take care of her for the rest of my life."

I was about to fall off this bar stool because I'm afraid of this man with his big hands, even without his shotgun. He could crush me with one of his hands.

He looked at me and said, "Lieutenant Wilbur, you seem like you are a pretty nice young fella. I really liked what you said about my Emily. I expect great things from her. She is the first of my children to go away to college. I'm going to take you at your word and I'm going to give you my blessing to marry Emily."

I almost collapsed. I certainly didn't correct him on my name.

He said, "Now, let's drink to that."

I said, "Sir, I don't drink."

He repeated, "Let's drink to that."

I said, "Yes, sir." I took a little sip of that bourbon and coughed. It burned my throat.

He patted me on my back and said, "You'll be okay, Son. I won't tell your mother."

We returned to his home, and he told his wife, Emily's stepmom, he had given his blessing for me to marry Emily. She said, "Welcome to the family."

I told him I'd like my biological father to meet him. I was on Cloud Nine. That was June 1963.

I wanted to get married in December 1963. Emily wanted to wait until the next year and go to the World's Fair. I said, "No, no, no. Let's set the wedding for Dec. 22." I wanted to get married as soon as possible because I feared losing her.

Another reason I was ready was I had saved a lot of money. I ate in the officers mess, so I didn't have any food expenses. My pay was good but if I got married, I'd get another extra $110 a month. My base pay as a second lieutenant was $232.30 a month. I had this big new car, and the payment was $105 a month so I had good transportation.

Our wedding was set for December 22, 1963. I asked my oldest brother, Alvin, God bless him, to come be my best man and my second brother to be one of the groomsmen and a fraternity brother, Horace Nelson, to be the third groomsman.

Emily was Episcopalian, which required me to take instructions in the Episcopal Church. I had been baptized in Mount Pleasant Baptist Church in Goulds, where I had attended Vacation Bible School. I had been active in Baptist Young People's Union. I had no idea what they did in the Episcopal Church, but ours was going to be an Episcopalian wedding.

I went to the main chapel at Fort Jackson, South Carolina, Chapel No. 1. The chaplain told me I would have three weeks of training. I went twice a week, for one hour a day. At the end of this three weeks, I'd get a certificate of completion and then I could go ahead and get married to my bride. I completed the instruction and called Emily that "all systems appear to be on go" for us to married. Emily made all the arrangements in Savannah, for our December 22 wedding. My car for some reason was acting up the wrong day, our wedding day! I had to keep it running outside of the church.

Father Matthew Caution was the Episcopal priest. I thought, *Oh, my goodness, all they do is say the Lord be with you and with thy spirit, let us pray. Stand up, kneel down, up and down.* I had never been in a church service like it before.

Here I was, my oldest brother on my right and with the other groomsmen, when Emily and her stepfather, David Singleton, entered the back of the church. She was in a full wedding gown, a veil cascaded to her elbows, a single strand of pearls, the only ornamentation needed.

I turned to my brother with a sudden instance of cold feet. "I don't know if I want to do this—this is for life."

"All you do is talk about this girl, how much you love her and how you want to marry her. Shut up, Wilbert," he said in a very low voice. Here came Emily down that aisle. When her stepfather lifted the veil, I lost all my fear. I thought, *Oh yes, this girl is the one that I want to marry.* With her hand tucked over my arm, I walked a little bit taller as we left the church.

After the ceremony, we left the church, got in the car with the motor running down and drove to Emily's mother's house for the reception. People wouldn't bother your car in those days.

We had a big wedding cake in the dining room. All the guests joined us cutting the cake and all those rituals. Our honeymoon was supposed to be in Atlanta at one of the nicest hotels. That notion was quickly erased because my oldest brother Alvin convinced me to follow him to Trenton for my new wife to see my mother. Alvin had a brand-new 1960 Impala, and that thing was faster than one of those Daytona 500 cars. He took off and left us.

We got as far at Fayetteville, North Carolina, when we hit an icy patch. The car spun around in the road two times. We were very lucky because there was little traffic at that time of night. It could have been a bad accident, possibly killed or hurt us badly. As luck would have it, we found one segregated motel with one room left. I had never driven in cold weather. I didn't even think to check the weather or anything before leaving Savannah.

The next morning I found icicles on my car. I had to use a pair of pliers to start my car by touching the alternator. Emily helped me. I started it up and we took off for New Jersey.

We had a minor mix up at our New Jersey home. My mother didn't have a Christmas tree. Emily always had a Christmas tree. She never didn't have one. She was a city girl. We country boys didn't have money for ornaments for a tree. We'd always cut our own tree in the woods and make our own decorations from crepe paper, nothing store-bought or fancy.

Emily got upset and she went upstairs, crying. She told me she wanted to go home. She'd never had a Christmas like this. I told her we'd get a Christmas tree. When I told my mother, she sent my younger brother to get a small one with lights already on it that plugged into the wall and blinked. That's what we had.

After Christmas, I took her back to Tallahassee because she was starting her last semester. I said, "Now that we are man and wife, I want to give you some money for any expenses you may have at school."

Emily, very proud, said, "I don't need any money. My father has given me all the money I needed. My father told me not to take money from any man."

"I have all this extra money I had saved from my per diem." I didn't have any bills, other than the ring that I got for her. I said, "You're going to have to take some money." I recall writing a check for the $225 and one for $75.

She said, "I don't need anymore. Don't send me any more money." Sure enough, she went back to school and that was the start of our married life.

My job was to go back to Fort Jackson, South Carolina, and find us a place to live after Emily graduated in April 1964.

LT BRYANT
BRIEFING SECTION
SWIFT STRIKE III

Chapter Five

Army Green

After graduation, while I was pursuing Emily, my Army career began. Even though my mother was in the hospital and living in New Jersey, I knew that I couldn't stay near her because there were no jobs. My Aunt Johnni and Uncle Thomas Williams invited me to come to Brooklyn until I received my orders for active duty.

I found a job at Jewish Chronic Disease Hospital. Because I was a college graduate, the director of maintenance had me as one of his 'gophers.' The first week there I got accustomed to all my tasks. I did everything including operating the elevator, checking his books, and checking his math in the office figures. He even offered me a position after my two years in the military.

I was primarily concerned when I would be called for active duty. Often I called my brother Bobby in Trenton to ask if there was any mail from the Department of the Army. About the third weekend I called and Bobby told me yes, there was a big brown envelope with the return address of Headquarters Department of the Army.

I zipped over to my mother's home. My hands were trembling as I opened the envelope. It read "You are hereby directed to report to Fort Benning, Georgia, to attend the Infantry Officer Basic Course, starting Oct 17, 1962 on active duty as a second lieutenant." That was great news. I still didn't believe that I was going on active duty. The first part was a six weeks Basic Infantry Officer Course with about 220 to 230 brand-new second lieutenants. Here I am thinking I was exceptional, and we had all these brown bar lieutenants. I was told to report on October 16. I took the Greyhound bus from Philadelphia to Columbus, Georgia, to the Infantry School. I took a footlocker with all my possessions, the same one I had taken to Florida A&M.

I arrived early, 7:30, while my report time was 9:00. At the bus station, I got in a cab and headed to Fort Benning. I'm a hurry-up-and-wait kind of guy. I was always advised by my professor of military science and my other military instructors to never be late in the military. I said, "Sir, Second Lieutenant Wilbert Bryant reporting for duty." The captain saluted me and asked to see a copy of my orders. He told me to step back, stand at attention and raise my right hand as he swore me in.

My base pay was $222.30 dollars a month, which was a lot of money because the most I had seen on checks was $30 from each of my four jobs at Florida A&M, for $120 a month. Next he gave me instructions on getting a gratuitous issue of fatigues, but as an officer I had to pay for my uniforms. I was to go to the quartermaster clothing store with this advanced per diem money to acquire two pair of boots, five to seven pair of fatigues, a field jacket and a couple of caps. I asked what was next for me and was told I wasn't needed until Monday morning when school started. I asked if I was free the whole weekend. He said yes, so I hurried down to Tallahassee to see Emily.

In late October 1962, classes would start. The mixing of races in that environment didn't bother me. ROTC Summer camp paved the way for me at Fort Benning, surrounded by Army Green.

They teamed the eight Black officers as roommates, which didn't faze me. I guessed they were trying to make us more comfortable. My roommate was Lt. Matthew Devoer of South Carolina State University, an HBCU. An outstanding army officer, he was the valedictorian in his high school, and also the number one in his ROTC class. He was from Gullah country, like Emily. This young man was as serious as a heart attack about studying, which suited me.

So Monday through Friday, we studied hard. We would study alone and then quiz each other. He was very loyal to his girlfriend Juanita at home. He never went anywhere after work ended at 6 o'clock on Friday. He stayed in our room, and he wrote his girlfriend a letter every single night. I'd never seen anything like that. If you're a single guy, you take off with about five guys and go to the movies together, you enjoy Columbus restaurants, maybe see the ladies there. Guys like to get out and have fun. Matthew never would go any place with us. I wanted to meet his girl and tell her that she had a really good guy. I attempted maybe two or three weeks to get him to go to the movies. He declined. His devotion to study made me almost mad because I'm thinking I'm a good student, but I'm not that good.

I thought, *I'm going to be myself and have a little fun from Friday night about 7 o'clock until Sunday about noon.* Then I'd go to a civilian Baptist church because that how I was raised. When I got back, I opened my book to begin my studies for Monday. I made sure that my boots were shined and all my uniforms were lined up, so I was ready.

We went through the basic course, and I did very well. I heard through the grapevine some officers had advanced knowledge of the exams. I didn't want any of that. Here I am studying and I'm trying to memorize things, using my good memory. Thank goodness my roommate and I were on the same page. We'd have to study for a three-hour exam the next day. I didn't need any assistance from cheating. One guy said "I know a guy who's got it." I said, "Don't tell me. I don't want to do anything like that." I was going to go in there based on what they taught me and what I had been able to memorize. I did well, scoring between 84 and 88 percent and a couple of times 92 percent.

About midway in the class, we were told about other courses we might want to take at Fort Benning. The options included airborne school, ranger school, and flight school in Alabama. I signed up for all three without knowing what they entailed. I always wanted to outdo my oldest brother so I signed up for airborne school because Alvin had, even though I had never flown before in my life. He didn't tell me I had to jump out of an airplane!

Airborne school was four weeks. The first week was called Ground Week, when you really got in shape. They taught the ins and outs of the parachute. The second week was the swing-landing trainer. We put a harness like a real parachute and jumped off a four-foot platform to practice landing on the ground. The third week we trained on a 30-foot tower where you jump out of the tower and slide down a guy-wire for about a hundred feet to a mound of dirt with a soldier on it to stop you. Little did I know I was about jump out of an actual airplane.

Upon completion of my weeks of training, they told us on Monday morning we were going to be involved in jump week, when we would actually jump out of an aircraft. I said, "What? My brother didn't tell me that." During those days we had limited telephone contact. I didn't even have a phone number for him to advise me on the actual jump. I was so mad. He should have told me I had to jump out of an airplane.

Next we went down to Fryer Army Air Field. Since I was a lieutenant, I was in the stick-leader position. I had to come out of the door first, with 22 men following. There was another second lieutenant on the other side with his 22 enlisted men. *Oh, my goodness.* I asked, "How high are we going to be?" I was told 2,000 feet.

We had practiced and trained. I knew what to do, but still I was scared. I couldn't tell anybody, certainly not those 22 enlisted men because I was one of their leaders. We took off in a C-118 or C-119 plane that had two propellers. We had certain procedures we followed before actually jumping out. When the jumpmaster said, "hook up," I hooked my cord on that steel line running down the center of the airplane. I looked outside. *Oh, my goodness, it is a long way down.* They taught in airborne school never to look down after exiting the airplane. Look at the horizon because if you look down once, the ground will be coming at you real fast.

Now there was a red and green light on the side of the plane. I stood in the door. My knees were literally shaking. The sergeant saw that. Every airborne student had a number on his helmet; I was 651. The sergeant asked me why were my knees shaking. I told him it was the wind from outside. He teased me, "You wouldn't lie to me, would you, sir?"

I was definitely afraid, but I answered, "No, Sergeant!" He told the guys we'd find out what it was like in a few seconds. We had been trained when that green light came on, we jumped. He tapped me on my rear end and shouted "651—Go." I jumped. We were supposed to count 1,000, 2,000, 3,000, 4,000. By the time we got to 4,000, the parachute would be open.

I jumped out of the plane and starting counting. The propeller blast wind hit me and flipped me over. I was only 155 pounds! I closed my eyes. That was another thing we were told—never close your eyes. It was very quiet, but something told me, *Open your eyes, dummy.* I looked up. Thank goodness that parachute was open. I thought, *this is easy.* I reached up and I grabbed my risers. I pointed my toes in a prepare-to-land position. *Easy,* I thought. We had been trained to roll over as soon as we hit the ground and to gather our parachute. I reached over and tried to get my parachute, but the winds were high that day.

I rolled up, but the wind got in my light parachute, threw me back down and dragged me about five feet. Again, I got up and grabbed my risers, but the wind threw me down again. Soldiers on the corner of the drop zone were laughing. One of the sergeants yelled for them to help that lieutenant out. They collapsed my parachute, and I rolled it up and put it in my bag. After this was all over, I thought, *This is like falling off a greasy log.* I couldn't wait to go jump again.

We had to complete five jumps to qualify to get our Airborne wings. On my fourth jump, we had all our equipment—food, C-rations, fatigues, all the things we would live off as light infantry men, in a bag strapped to our bodies. Once we jumped out of the aircraft, we hit a quick release to drop our bag before we hit the ground.

Unfortunately, my parachute was twisted, a cigarette roll malfunction, a major problem. The worst—if I didn't unravel it, the plunging landing could have killed or badly injured me. We had been trained to get out of that situation. If your risers are twisted, extend both arms and snap your body in the opposite direction of the twist. I did and the parachute opened with everything now normal. I hit the quick release, and my bag dropped. Upon landing, I rolled over, gathered my parachute and ran off the drop zone.

Our instructor told us if all went well, we would have our final jump tomorrow and we'd have our graduation ceremony at 1,400 hours, or 2 p.m. We went up, I jumped and counted 1,000, 2,000, 3,000, 4,000, the parachute opened and I landed and rolled it up. It was all over. Graduation took place, and I received my wings. I was airborne qualified. You couldn't tell me I wasn't Superman after those five challenges.

At Fort Benning, I also had signed up for ranger School, which in my view was the toughest training an individual could get. It was nine weeks long, with a jungle phase, a swamp-water phase, and a mountain phase. In Georgia it was freezing cold in February. There was no way I was going to jump in water when it was 25 to 30 degrees. In the battalion headquarters I knew the assistant adjutant. I asked him to please take me off those orders. I offered to come back in the summer when it was warm and take the training. He told me my name was on those orders and he couldn't change them. To make a long story short, I paid him $25 to please take my name off these orders. He took my name off, warning me to not tell anybody about what he did. I told him I would never do that. So I didn't go to ranger school.

Next, I took the exam for flight school. I wanted to fly a helicopter. I passed the written portion of the aptitude test. Next, I had a physical exam. The doctor told me I had a heart murmur which disqualified me.

I argued I could not have a heart murmur. I'd played varsity basketball in high school and I'd played on a state championship baseball team with never a problem. The doctor turned me down for flight school, saying I could reapply the next year. During those times only a handful of Blacks were admitted to flight school. I didn't know that at the time, but as I look back and reflect on it, that's exactly what happened.

I stayed around Fort Benning on temporary duty status until March 31. I received orders assigning me to the USATC Fort Jackson, South Carolina, to report on April 1, 1963. I was assigned to the Army Training Center where the new recruits come into the Army. I would be involved in training them. I really didn't care for the assignment because I was an infantry officer, and I wanted an infantry division.

I called my brother Alvin who told me, "Don't worry about that," he said. "Over time this will come. Do your very best in whatever job they give you." He reminded me I was being rated in everything that I did. This was to be my first teaching experience.

When I started at Fort Jackson, I had a few butterflies in my stomach because it was the unknown. It was the first time that I was on my own and having to find my way without someone telling me where to go.

At Fort Jackson I made good friends with Earl Williams from Florida A&M University class of 1961, and 1st Lt. Willie Laster, an officer from Tuskegee Institute. They looked out for me and guided me on what to expect there. Two other officers were good pals but from other backgrounds. They were my floormates, next door in the bachelor officers quarters. Another one was 1st Lt. Frankie Lockchiss from Brooklyn, New York, a young white officer, very friendly. When he left a few months later, I got another floormate, 2nd Lt. Augosto Caesar Sanchez from Puerto Rico, a graduate of the University of Puerto Rico. He was an outstanding army officer. All these guys were wonderful friends who guided me.

My first month at Fort Jackson, I was the executive officer of a basic combat training company of about 220 new enlistees. I did whatever the company commander assigned. Little did I know how closely I was being observed.

The first month as executive officer, I had a company commander, Lt. George Pockhaber, a Midwesterner. He became an officer through OCS or Officer Candidate School. I was ROTC or Reserve Officer Training Corp. The OCS officers, in my view, had little respect for ROTC officers. The OCS guys claimed we walked in the Army easily, getting our college education, while they thought they had to work their butts off before any kind of college education. He ordered me to take the company out and get exercise in the morning. He wanted me to run them. I answered, "No problem, sir."

Before breakfast, we were running one morning and here comes Lt. Pockhaber driving into work. Later he told me that whenever I saw his car coming up the road, I was to bring the company to quicktime and render, run in place, and salute him. He was mentally punishing me because I was ROTC. I didn't argue, just agreed. I must have done well because I was called up to brigade headquarters and I was selected for another bigger task.

There was a major training exercise in the United States called Swift Strike III (SS3). The orders came down: 2nd Lt. Wilbert Bryant was going to be sent away for one month to serve as a briefing officer in the Joint Visitors Bureau for Swift Strike III headquarters in Spartanburg, South Carolina. My superiors had observed me in front of my troops and concluded I spoke with authority and could give twice daily updates to a group of senior officers. I was teamed up with a young Air Force first lieutenant. I believe I was selected again for my speaking ability and my ability to memorize. The first time facing all those generals and colonels, I mentally blinked until I grabbed hold of myself. This was not like training or being sports director but so much more. I reminded myself, at Florida A&M, I had been taught how to give a briefing. So I plunged in and briefed those generals and colonels without a hitch. When I came back to my unit, my leaders at Fort Jackson had been told I did an outstanding job at Swift Strike III.

Next I was reassigned from the BCT company and became Officer In Charge of a close combat range at Metz Range. This is an exercise where 10 men walk in line, firing at targets that appear in front of them. First the men would practice and then there would be a live course, firing live ammunition. I had been warned nobody could get hurt. I had to teach these soldiers to always keep their weapons pointed down range. Never point to the left or to the right because with their hands on the triggers, they could easily shoot the soldier next to them. If that happened, the OIC could lose his commission, be kicked out of the Army, and go to jail.

I was the primary instructor of the close combat range. When a company came to my location, I taught them a block of 50 minutes instruction. I had a script of 12 pages of information which I committed to memory. The script was supported by the hidden assistant instructors. I'd say a certain word and they'd jump up and fire. It was all choreographed. In my opening remarks, on cue, soldiers hidden underground would rise up shouting; it really got the soldiers' attention.

I used this phrasing: "Stay alert and stay alive out here. Camouflage is being used and there are enemy soldiers all around you." I must have spent two weeks memorizing the information. I practiced what I had memorized like I had done in college.

To teach the class, I had a microphone on my chest and I used great deliberation and energy. A number of training inspectors witnessed my first block of instruction, taking notes on their clipboards. My brother had told me to act like they weren't there because I was in charge. I was a little nervous with soldiers' lives at stake, but once I started talking, I had it down pat.

In April 1964, after Emily graduated, it was time for us to set up our first home. Emily's professor, Dr. Carriemae Marquess, had a friend who owned an apartment building close to Columbia's Allen University campus. Emily's professor instructed me to see her friend who would rent us an apartment. It was a one-bedroom fully furnished apartment, with a living room, a dining room, a bathroom, a bedroom, and a small kitchen, all for $50 a month.

In April 1964, I went to Emily's graduation and picked her up in Tallahassee. She had a huge trunk for all of her things. When I'd left Trenton to go to Fort Benning, all I had was a footlocker. I certainly learned girls had more clothes than boys.

In Columbia, I had painted the apartment and I thought I had done an excellent job. One of the first things Emily noticed when she got there was my paint job.

"We're going to have to paint this apartment." When I told her I just painted it, she said, "Oh no, this is not good enough."

We repainted the apartment and got it all squared away up to her standards.

Emily would drive me to work each morning and would come back later to pick me up. Little did I know, Emily had gone out, after maybe a month, and found a part-time job in a department store in downtown Columbia.

When she told me about this part-time job, I told her I didn't want her to work. I had promised her dad I would take care of her.

She replied, "I've always taken care of myself, and I've always had a job. It is only part-time, and I really need to do something. I can't just sit in this apartment day in and day out."

I didn't argue with her.

Those first few months were very interesting getting settled. We didn't have a lot of extra money, so I imagine that's why she took this job downtown. At that time, my base pay was still $220.30 a month plus $110 basic allowance for quarters, but Emily's money—that was a great addition.

Everything was going smoothly. As OIC of the close combat range, I performed well. I was moved again and made a company commander which means I would be in charge of approximately 220 new enlistees. My company was Company C, 4th Battalion Training, 1st Training Brigade.

By then I had been promoted to first lieutenant. At that point, I was still trying to impress Emily. One day I had the post duty officer detail, which meant my company provided guards in front of the post exchange, the bank and the post office—all key facilities on Fort Jackson. I played it up to Emily, telling her I was the top guy running the post that night. When everybody goes home after 6 p.m., I told her I'd be in charge of the post.

At that time, Emily would bring me my dinner when I had evening duty. Thinking that HQ would be empty I asked her to bring my dinner to regimental headquarters and leave it with the NCO. Never once did I think that the regimental commander, a colonel, would still be there. The colonel was working late that night along with his adjutant. Emily pulled up at the headquarters—I heard all this later. There was a sign over the door for the "Colonel's Entrance." Emily opened the door and walked straight in. Who was looking directly at her? The colonel himself, eagle rank on his shoulders. He wished her a good evening and asked if he could help her.

"My husband is Lieutenant Bryant and he is the acting colonel tonight," Emily said. "I'm bringing in his dinner. Is he here?" The colonel told her I was out checking the troops. She said, "Would you give my husband his dinner?"

Oh no! Why did she do something like that! A little first lieutenant's wife giving the colonel his dinner! Immediately, I received a call on the radio from the adjutant telling me to "get up here right now!" I dashed up there. I thought maybe it was a fire or somebody had broken into the safe.

Capt. Israel Lopez was the adjutant. I will never forget his name. He yelled at me for about 10 minutes.

"Don't you do this again. Having your wife to come up and ask the colonel to give you your dinner? Are you crazy?" He was right up in my face. Almost like he was spitting he was so mad. He chewed me up and down. "Do you understand me!"

"Yes sir, it will never happen again."

He snapped back with "You're right. It never will. Get outta my face!" So I really messed up.

When I got home, I told Emily not to ever do that again. She said,

"You told me you were the acting colonel." She had done exactly what I told her to do. Thankfully it blew over.

Emily joined the Brigade Officers Wives Club. She has always been outgoing, active and very friendly. In Columbia, South Carolina, activities still were segregated outside the post. Emily signed up for a horseback riding event with the wives. Any wife who wanted to sign up for horseback riding, chess, or knitting did so. The event came and went. No one called Emily to let her know about it.

The next meeting Emily stood up to complain. The brigade colonel's wife was the president. Emily told her she had signed up for horseback riding and nobody called. The colonel's wife apologized and said she'd look into that.

When Emily got home and told me what happened, I said "Oh no, you didn't! You know that they're not going to invite you to the event in segregated Columbia. You are not going to go to these places."

Emily replied, "Well, they should not have offered it then." I hoped it was not a big problem, only a minor incident, but I think it spurred them on to give me another assignment.

By this time Emily was expecting our son soon. He was born, coming into the world, kicking, January 13, 1965. I was the happiest man alive. When I went back to the apartment, I realized I didn't know whether the child was a boy or girl. As I recall I called the hospital and asked what did the Bryants have? The nurse told me I had a healthy son. *Oh, my goodness!* When I went back the next day, they wheeled him out to the plate glass window in the nursery. Such a beautiful baby boy with black curly hair. So that was the start of the three of us.

Shortly thereafter, I was placed on orders assigning me to Germany. I always wondered if that incident about the horseback riding had anything to do with that because it was right after that event I came out on orders.

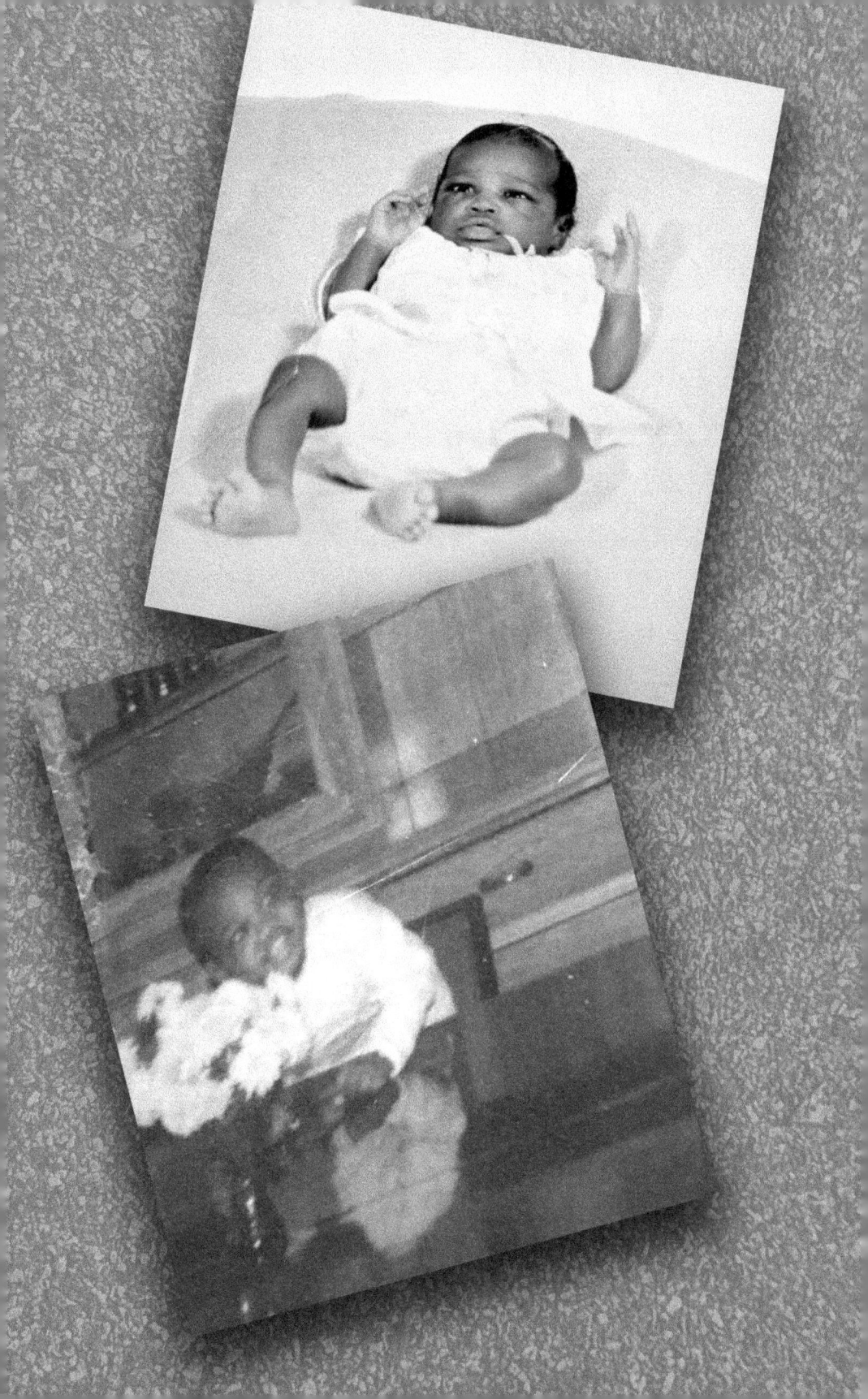

Chapter Six

German Immersion

My orders were for Würzburg, Germany. My move had to be quick, so I had to send Emily and my son Wil Jr. to Savannah, an hour and a half drive away from Columbia. Then I had to come back to clear the company books and check out of Fort Jackson. All blankets, sheets, pillows, and equipment and property had to be accounted for before I could get a release from the officer replacing me. Unfortunately, I didn't have enough rank to entitle my wife and son to travel with me; they would have to come later.

The Army's orders put me on a Military Sea Transportation Service ship, the USNS *Buckner*. It departed Brooklyn Army Terminal for Bremerhaven, Germany on February 20. Ten days on the Atlantic Ocean during the winter was very cold—I had never been on the water for that amount of time. I thought of the great Titanic. I reminded myself that this was the Army. There were some 1,200 to 1,300 troops and dependents on this transport ship; it would be safe.

On board, I was assigned to a table with a young white wife and her three children to help her. She had a baby in her arms, and she had two very active boys, maybe three and five years old. They cried and carried on, but the boys quickly took a real liking to me. She was on her way to join her husband who was a first lieutenant in Germany. I sat with this young family for breakfast, lunch and dinner. The boys and I got a 10x14 sheet of paper and every day we put dots on it to show our progress across the Atlantic Ocean.

We did daily fire safety drills, which meant we all had to leave our cabins and be on the deck of the ship right behind the lifeboats to simulate an emergency if we had to evacuate the ship. The first day I made a mistake by getting out there close to the start time, which put me close to the railing on the ship, this huge troop transport. You could feel the ship moving. Oh, my goodness. I couldn't stand it. So, if the fire drill was scheduled for 10 a.m., I got there at 9:30 a.m. so I was against the bulkhead. I could swim, but I knew if I went overboard into that water, I was done for.

On the ninth day when we hit the English Channel and the Dover straits, the seas got super rough. The ship pitched up and down, left and right, so bad we had to put safety rails up in our bunks at night. I didn't sleep any that night, and the other three officers in my cabin were lined up go to the bathroom. I was determined that wasn't going to happen to me. I took the pills given to us and they helped. I tried to think of anything other than being on that ship. I thought, Oh God, you get me off the ship, I promise I will never get on another ship in my life. Finally we pulled into Bremerhaven, Germany.

When we offloaded the ship, it was a somber sentimental departure from those little boys. I knew that I was going to love being a father to Wil Jr., and I was so sad because he and Emily weren't with me. The boys' mother asked me to meet her husband, so he could thank me because her two little boys had loved me and they listened to me.

When I went down off that big black deck on the ship, I got down and kissed the ground. We met her husband. I shook his hand and told him he had a wonderful family and I was happy to help.

I boarded a troop train which took me to Würzburg, Germany. My sponsor picked me up and took me to the bachelor officers quarters on Würzburg Post. I signed in the unit and the next day, went to meet the colonel, the post commander. I was eager for an Infantry assignment.

I found out I was assigned to Würzburg Post, which was an alternate specialty for an Infantry officer. I really didn't like that. However I remembered what my oldest brother told me earlier—regardless of what job you have, learn the job and do it very well every day. My job was assistant military family housing officer. I feared this was a career-ending job, but it didn't work out that way.

I eagerly awaited Emily and Wil Jr. I went out in the local economy to find an apartment for my family to live in because I didn't have enough rank to live on the post.

I found a brand-new apartment in a little town called Gerbrunn, right off the post. I got first looks at it because I worked in the military family housing office. They hadn't even posted it; they let me and Emily have it. Two-bedroom, one bath, living room, dining room and a kitchen. Brand-new! The furniture came from the military family housing inventory—new couch, new carpet, new baby bed—everything we needed.

In June 1965, Emily and young Wil arrived at Rheim Main Airport in Frankfurt, Germany. It felt like the second happiest day of my life because I hadn't seen my son in five months. They flew the Pan Am 707 jet out of Charleston, South Carolina, nonstop for Frankfurt, West Germany.

I hoisted Wil Jr. high in the air and marveled at how much he'd grown, his round, pudgy baby muscles had him already crawling. Somehow I didn't scare him—probably Emily had prepared him with photos and stories about his dad.

I had my Chevy Impala now so I picked them up from the airport. My sponsor warned me about driving on the Autobahn where there were no speed limits. In my first experience, I got in the left lane to pass another vehicle. I was doing about 75 miles per hour. I saw some lights flash way behind me. I didn't pay any attention. I took my time passing. When I looked in my rear view mirror that car was right on me. Beep! I never got in the left lane again unless I drove it really fast. They drove crazy, maybe 100 miles per hour, though everything was in kilometers. My Chevy didn't compare to the BMWs or the big Mercedes-Benz sedans.

Emily fixed our apartment so nicely. We had a little group of German friends that lived nearby, Hans and Barbara Kloters with their little baby son and Wolfgang Ruckman and his wife. Emily loved going out in the village and shopping. Emily had studied French when she was in high school, and I think she took French in college but I wanted us to know enough German for her shopping.

I didn't know any German so I immediately enrolled in a German class on the post. The course was a satellite of the University of Maryland. There were 15 to 17 students in the class. The first night in this class, the professor walked in and said, "Guten Abend. This is the only night that we will speak English in this classroom, so I recommend, if you are not serious about this class—you'll have a lot of homework—I recommend you drop the class." We had class twice a week.

The second class he gave us several exercises and our homework. He said, "Guten Abend meine Damen und Herren." I memorized phrases and individual words. By the third class, over half the class had dropped out. I was one of the six or seven left in the class. I learned to speak a little German, enough to get by, especially how to count the money. The Deutschmark (DM) was the currency in use in West Germany at the time.

I explained to Emily we got about 3.60DM to every dollar. I warned her to be careful; if an item was priced at 9DM, she shouldn't be giving them more than three dollars. You should get change back. The German merchants loved American dollars. At the American Express on the post we changed over some money from dollars to Deutschmarks. I reminded her she didn't speak German and they might shortchange her. She told me she wouldn't let them.

When I was promoted to captain, Emily participated in the promotion ceremony. She pinned on one of my leaves. The colonel pinned the other one on. I think she's been there for almost every promotion that I have received in the military except one, and that was the first lieutenant promotion.

We lived maybe a half-mile from my job. Every day I'd walk up and leave the car with Emily because she liked to run around all over Germany.

We had two very good Army friends we made shortly after Emily arrived. One was a white officer from Alabama, 1st Lt. David Wright and his wife Jean. They had a child like us. He was my best friend when we were stationed there together. We socialized and did everything together. Germany has several celebrations throughout the year, Fasching and Kiliani wine tasting. For Fasching, there were parades with floats, everybody in funny costumes and masks. We sampled the sweet treats with our friends, including Berliner, a doughnut, Krapfen, a pastry, and Schmalzkuchen, a fried pastry. I really liked the Krapfen. We even got dressed up in costumes as an Indian chief and his woman. We wanted to do what the Germans did.

We stayed there for a good while, but Vietnam was escalating and many officers were being taken from the post and sent to Vietnam, which meant that the officers in my unit had to take on additional assignments because the Army was not sending replacements.

Now I was an infantry officer, and I knew the combat clock was ticking for me. I never told Emily or anyone that I wanted to go to Vietnam. I never mentioned it to the Army because a lot of people were getting hurt and killed. Deep down inside I was an infantry officer. I wanted to do my part to help defend our way of life in America. Vietnam was the only war going on, and I wanted to be a part of it.

In late '65, I was given an additional job as the commissary officer. Commissary manages the food. That operation was a humongous job. Headquarters told me as an infantry captain, my job for family housing had been outstanding so I could take on all these extra jobs. That commissary job gave me gray hair. All the fresh food came in to serve the entire community of the 3rd Infantry Division Headquarters in Würzburg. A major general commanded the division. His entire staff was there, and we were in the support of several thousand troops who used the commissary. I had to make sure it was fully stocked. I really didn't want to do this every day. They kept piling jobs on me.

My silent wish came through. I was called into the colonel's office. He said, "We have orders from the Department of the Army that say you will be assigned to Vietnam."

This was big news. I asked, "When is my reporting date?"

"Your last day here in Germany will be August 31."

The army gave me 30 days to get Emily and my son back to the United States and get them settled. Then they would ship me out to Vietnam right after that.

On the first of September we flew into McGuire Air Force Base outside Trenton, close to where my mother lived because that's what was listed as my home of record in my military records.

I had a 30-day leave. We stayed with my family for a few days and went down to Savannah to see Emily's mom and dad. It was a very emotional time for me because I didn't know whether I was going make it back home, considering the fact I watched all the press reports, all the casualties in Vietnam. KIA—killed in action. I departed the United States on October 2, 1966 heading to South Vietnam, arriving October 3.

Chapter Seven

Vietnam: Leadership Under Fire

When I got off that plane in Saigon, humidity swamped my face and hands and arms. The air felt damp, almost fetid, not like Florida's sunny breezes and rich soil smells. The landscape was so different from anything I had ever known.

All of a sudden, I heard those B-52 bombers dropping huge 2,000 to 3,000-pound bombs in Cambodia and Laos, 30, 40, 50, 60 miles away. I was afraid, but I kept trying to think if others have been here and they made it back, there was a chance I could. With all my might, I tried to keep thinking positive.

We boarded buses that had wire mesh over the windows so nothing could be thrown inside. Many times, the Viet Cong threw a hand grenade in the window to kill everybody. I hoped we could get there without an explosion in my first hours in the country. Some buildings were open to the air, the front wall having been blown off.

As we entered central Saigon, thousands of people crossed the streets on bicycles, in rickshaws, on little motorcycles, and in small buses and trucks. In this large city, I couldn't believe the ox-drawn carts. It was an old city, with French influence in the more modern buildings. Music poured out the window of little cafes. Most of the people wore conical hats woven, I was told, of palm. The women were nearly all in white high-collared tunics over loose slacks, except the ones in '60s attire who whistled and waved to us, the American GIs. I was shocked—there was a war going on around them; we could hear the bombs all through Saigon—yet the Vietnamese acted like nothing was wrong.

At our temporary lodging in the Rex, a French-built luxury hotel, journalists wearing press passes roamed the lobby. The U.S. military command gave its daily briefings from the rooftop terrace bar. I wouldn't let them put me on the second or third floor because I feared we'd be attacked by Viet Cong in the night. I didn't have a weapon. At my first opportunity, I asked when they were going to give us a weapon so we could defend ourselves. I was told I didn't need a weapon.

On the top fancy floor of the hotel, we were fed hot dogs, beans, a vegetable and a piece of bread. We had that meal twice in one day. Some luxury. They issued us several sets of jungle fatigues made for operations in a jungle environment. I had pockets from my ankles all the way up to my shoulders.

Again, I raised my hand. "When are you going to give us a weapon?" They told me I was pretty safe here and not to worry. I didn't agree but I piped down. They told us the next day at 10 o'clock we would be given our assignments.

South Vietnam's defense was divided into four corp areas. I-Corp was the northernmost core up near the demilitarized zone separating

South Vietnam from North Vietnam. The Chinese supported North Vietnam. Next was II-Corp, III-Corp and IV-Corp, which was south of Saigon. It was called the Mekong Delta, and it was loaded with rivers with outlets, boats, and water. I didn't want to be assigned there, anywhere but there.

The next day in meeting room, they started to call our names. I don't think I slept two hours the night before. When my name was called, my hands were shaking as I opened my orders; I was assigned to I-Corp, way up north.

One of the captains said, "You will be assigned as a senior advisor to a Vietnamese infantry battalion."

I answered, "I'm just a captain."

"Your training is far superior to any training that they have ever received. It's a battalion of 400 men, operating just south of the demilitarized zone," he said. "You have to worry about rockets and mortar fire because you will be fighting against the Viet Cong. And snipers and booby traps."

Before I arrived, I had read Bernard Fall's 1961 book, *Streets without Joy: the Military Debate in Indochina*. He wrote about booby traps and mines—those were what I feared most. I was told to be ready tomorrow morning at 0800.

We were loaded onto a C-130 aircraft, painted camouflage to blend in with the terrain. A C-130 was the workhorse of the Army and Air Force, but taking off and landing it sounded like a mad bull. We flew from Saigon to Da Nang Air Force Base, which was the largest Air Force base in the world at the time. They told us it was relatively safe because the South China Sea was off to the right. We circled out over the South China Sea and flew at an altitude of about 1,100 feet into Da Nang. When the pilots reversed the engines to land, it sounded like it was crashing. I was a trained paratrooper and

I wished for a parachute. I found out later it was the plane's propellers that were so loud.

We landed. I was sweating with the heat and fear, even more so when I saw stacks and stacks of silver caskets, thousands, next to a C-141 cargo ship to fly the KIA's home to the U.S. I wanted to charge back on the C-130 and get out of there

A major I had met took me aside and advised me, "Take it day by day. Don't look too far out into the future if you want to survive here. Make it through one day at a time. Mark it off at midnight every night and focus on the next day. If you look too far ahead, you will end up getting killed." I followed his advice for the next six months.

Finally they issued us our weapons. I was given a rifle and as much ammunition as I wanted. We could also ask for more weapons at our units. I had two weapon pouches and I had two magazines each with 20 rounds. I loaded one in my rifle. My safety was on, so I didn't have to worry about it inside the plane. I had four hand grenades. Depending on the operation I would carry an M-70 grenade launcher and a bandolier of those grenades strapped across my body.

The next day they flew us into Hue, South Vietnam, or the old Imperial City, to the headquarters of the First Army of Vietnam. I was assigned as senior advisor to Advisory Team Three. I'd be operating in the Hue, Qung Tri, and Dong Ha area about 10 miles in from the South China Sea.

I was told my counterpart was a Vietnamese major. I wondered how could this be. I was just a young captain, and I would be advising a Vietnamese infantry battalion of 430 men. I quickly realized they were poorly equipped. I gathered my gear, all in one duffel bag.

They put us on South Vietnamese helicopters, flying out officers and NCOs every five minutes. I'd heard all kinds of rumors about them. I prayed to make it. When they were ready to land the helicopter, the ground crew popped a flare, a sign to identify their location and that it was safe to land. The pilot had to identify the color of the flare to know if it was a friendly flare. I disembarked and joined my three American advisors.

I hid my nerves and my fear. I figured I was a former athlete; I was in good physical condition; I could run if I needed to. I said to myself, *I can make it out,* again, using a drill of mind over matter.

At that point, I didn't trust anybody, not any South Vietnamese. They have warned us that some of the South Vietnamese behaved friendly during the day but at night they were really Viet Cong sympathizers.

I was the senior American advisor on this team. I had a young White first lieutenant who was my assistant and two noncommissioned officers to help me. My master sergeant, a Filipino Black sergeant, E5, was a big guy. My first duty or responsibility was to introduce myself and let my guys know exactly what I expected.

They neglected to tell me a key fact until I got to headquarters—the advisory team before me had been captured and killed by Viet Cong. Their bodies never found. All headquarters said was to be careful. I said, "What?" They warned me to keep my eyes open. My morale crashed.

At night, they assigned two bodyguards for me. I told my counterpart, who spoke good English, to tell them to stay at least 15 yards away at night. I instructed him to tell them not come up to me at night because I would have my weapon with my safety off, and I would shoot anybody close.

I said to him, "Tell them right now." He told them. I asked him, "Did you ask if they understood?" He did. I took all these precautions because I wanted to get home to my wife and son.

Those first few days were very tough. We were about to move our position when a sniper about 50 yards away in one of the tree lines shot my radio telephone operator, who was standing next to me, right between the eyes. I was spattered with his blood. My team knocked me down. My RTO died instantly.

I saw myself through the sniper's eyes. Through his high-power scope, I was taller, with officer's insignia, and I was Black. I had been talking on the radio, which was on my RTO's back. It was my first time coming under fire, and I had been there two days.

How to prevent horror from happening to me? I decided to wear my steel helmet all the time. I even shaved with it on. I had my rifle with me all day and at night. I ate lunch with it, locked and loaded with the safety off.

We went on operation every third day fighting the Viet Cong, sometimes in water up to my chest. I had to carry my rifle over my head so it wouldn't get wet. I had all the equipment to oil it and clean it, but I wanted to avoid that.

At night I didn't sleep. I had a silk hammock that I kept in one of the many pockets in my jungle fatigues. On the off day I'd string the hammock between two trees and I would sleep with my officers around me. I would close my eyes for four or five hours.

I washed my clothes in a little creek. I had a bar of soap wrapped up in a plastic container in one of my pockets. I'd lay my clothes on the bank and in 10 minutes they'd be dry it was so hot.

Every night we dug me a foxhole slightly below the ground. If we got hit with mortar fire or rocket fire, only a direct hit of shrapnel would get me. I could hear the vibration of a hit, but none of that metal would hit me below the ground level. My profile wasn't big so they didn't have to dig down far. I had a houseboy who would blow up my air mattress and put it down there. I kept my rifle on my shoulder and my stock across my body with my safety off. That's how I stayed ready. In the mornings around daybreak, I'd close my eyes for about a couple of hours because the whole battalion would be getting up around me.

Around 11:30 every day, wherever we were, even if we were in a shooting war, the enemy would stop fighting. On both sides, the mess teams with their black pots came up and started boiling their fish and rice in the field. We lived off the land. Our mess teams had thrown a hand grenade in a school of fish. Next, they cleaned the fish and boiled it and that's what we ate every day—rice and fish.

Every 14 or 15 days we'd come in from the field and stay in the small American compound in the old Imperial city of Hue. A one-quarter-ton Jeep would pick up me and my three American advisers and take us in. We'd stay overnight and catch a movie in the compound. In it, sandbags covered everything in case we got hit by mortar fire. At the compound I was offered American food, but I found out I couldn't eat it. I got diarrhea because I was accustomed to eating rice and fish. Many times, we'd travel through small villages with trees of small, sweet bananas. I asked my houseboy to cut a half a stalk—a luxury food, a banana.

The next day we had to be back at our unit by 3 p.m. because that's when the Viet Cong and the NVA soldiers started to ambush along the road. Our Jeep would stop at a precoordinated position. A platoon or a company would be hiding in the bushes, waiting for us. They would pop a flare and let us know that they were there. They'd rise up and we'd go back to our unit. We had that break on a 14-day cycle.

As the senior advisor for the battalion, I was near the headquarters, so I figured the Viet Cong would have to get all those guys before they got me if we were ambushed.

We Americans had a mail drop every 14 or 15 days. All I wanted was mail from Emily. She wrote every other day. I would not have made it out of Vietnam were it not for her. Emily would send Polaroid photos of my son and herself. The church and another friend sent me biblical readings. I got letters from my friends in Germany. I read those every day. I kept them wrapped in plastic in one of my pockets. When we got the mail, I'd organized my stack of Emily's letters by the postmark date. I'd take the first message that she wrote and read it first and then the rest in the order they were written.

Daily we had a small spotter aircraft circling over the South China Sea. Any time we had a sighting of enemy location, I would call the pilot on my radio frequency to tell him where the enemy soldiers were. Overhead he could see their location and he'd give me the coordinates. He would contact the Air Force F-105 Thunder Chiefs, circling out over the South China sea about 50 to 60 miles away. They would fly in and drop their bombs, napalm or whatever it was on the enemy position. We'd go in and mop it up. The Viet Cong and the NVA would scurry to drag out all their dead and wounded. Bodies and men dying around me was disturbing, but I had to get acclimated to survive.

For my team, the four Americans, I had an escape plan. If we got surrounded, we were going to make a mad dash to the South China Sea. I had sent my sergeant out and had him cut four bamboo sticks a foot long. The stick had a narrow circumference, a tube. I tucked it inside my uniform. I didn't let any of my South Vietnamese counterparts know. If we got surrounded, we'd run straight into the water, lie down flat on our backs, breathe through the bamboo tubes, and lie there until the enemy soldiers left. They wouldn't stay in a position very long because the aircraft would come bomb them.

After 30 days I believed I was going to make it. If they got me, I planned to get about 100 of them first. That's how I felt as a young 26-year-old officer.

Emily wrote me, when I had been over there about two months, she had not seen one military allotment. Not only did I draw base pay, but also overseas pay and combat pay. All that money was untaxed. I wanted Emily to have the great majority of that money. I didn't need any money in Vietnam because I was in the field fighting. I put 10 percent in savings. The situation made me furious, so I got my PRC-25 radio, called headquarters and told the boss I needed to come in.

At headquarters I raised holy heck with the personnel people. I yelled, "My wife has not received any money whatsoever. I need to get this straightened out." They apologized and said they'd get right on this. I said that was not good enough. I wanted it done while I was there in front of them. Because when that helicopter came to pick me up, I'd be gone. I demanded to know exactly when my wife was going to get the money. They were able to assure me she would get it within seven days.

During that time, the military had a system in place to keep the morale of the soldiers up called the MARS system, the Military Affiliate Radio System. In the compound, soldiers could place a call to their wife or loved one and talk for three minutes. There was always a long line, so I had to get there early. I was anxious to hear Emily's voice because it would inspire me and boost my confidence. Emily was not accustomed to the military lingo. We had to say *over* at the end of every remark. "Can you hear me, okay? Over."

After a couple of tries, she got it right and we were able to talk. I asked if she was getting her allotment—over. She said she was getting the allotment—over. The organizers interrupted "Captain, your time is up."

Their plan was for me to fight in the shooting war for six months. Six months went by in a hurry. But holidays were hard. At Thanksgiving, they brought us in from the field to eat a full Thanksgiving dinner. We had to be back in our Jeeps and go to the field before 3 p.m. On Christmas Day, it was the same, but it was very sad because Christmas music played in the compound. We called those guys who lived in it "compound warriors" because they had a nice cushy job. They weren't living under the stars. Even though those holidays were the most depressing times for me, I told myself over and over again, I'm gonna make it out and spend the next year's holidays with my wife, my son, my mother, and my brothers.

About the time scheduled for me to come out of the field, I was so relieved. The colonel told me I was to be the senior advisor at a Vietnamese training center just outside of the old imperial city of Hue, which meant I would sleep in the compound every night and not out in the field.

A major enemy attack north of our position near the Demilitarized Zone killed several advisors. After having been in this new job one week, I got a call phone from a major who worked in the 1st ARVN Division Headquarters. He told me Col. Kelly needed me in a new job. He wanted me to be the senior briefing officer in the Division Forward Tactical Operations Center. He told me I'd be in the field but I'd be sort of protected.

I replied, "Look, Major, I have served my time in the field, my six months. It's time for someone else to do jobs like that. Tell the colonel I do not want to go out there. I've been told I'd have six months in the field and then work in the headquarters compound. I intend to stay."

About an hour after the major's call, I received a call from the colonel himself. He said, "Bryant, get your duffel bag packed. I will have a helicopter out there in one hour. Be at the airstrip. Report to me."

All I could say was "Yes, sir."

The colonel said, "You are a survivor. You also know what's going on in this section. I need a combat-trained officer to provide leadership and keep track of all the battalions operating in the I-Corp area. There will be two of you assigned. You will be my primary briefing officer. I've been told how good a speaker you are."

Just like all those earlier times my skills propelled me to a position. I said, "But sir, with all due respect I was told..."

He cut me off, "Forget what you were told. This is what I want you to do."

I just repeated, "Yes, sir."

Here came the helicopter right on schedule and picked me up and took me to Division Headquarters. My work day began at 6 a.m. and ended at 6 p.m. I worked this cycle for the next six months. I had a briefing at 9 a.m. and at 4 p.m. to all these visiting dignitaries who came to the division headquarters.

We lived in a GP medium tent erected just before dark. We knew that the Viet Cong and NVA were monitoring us through their binoculars. We wanted them to see us setting it up. When darkness folded over us, we knocked the tent down, folded it and moved it to a different position 1,000 yards or more away.

On a good night, we'd get maybe four hours of sleep. Usually I would only sleep maybe two hours. Although we had a lot of guards around us, I did not trust anybody other than myself with my security. I needed to survive another six months.

About halfway through my assignment in Vietnam, I was eligible for R&R, Rest and Recreation. A soldier could meet his wife or loved one anywhere in 12 to 14 countries. You could request Hawaii, south Australia, Taiwan, you name it, and you could go. Emily and I agreed to meet in Waikiki Beach, Honolulu, Hawaii. My R&R was slated for April 1967. Emily and Wil Jr. came over and we spent seven days together. We were given a nice cottage on the beach. I was so happy to see my little family.

When she saw me, Emily started crying.

"What's wrong?"

She said, "You've lost so much weight."

"I have not lost that much."

She insisted I had. I told her about the rice and fish I ate every day. She had to get accustomed to me again because I was skinny as a pencil. Our R&R trip was just wonderful.

When that charter aircraft came back, there were 300 soldiers going back to South Vietnam. I can truthfully say there was not a dry eye on that aircraft. Every soldier cried, and I cried, too. I didn't cry in front of Emily and Wil. I was so afraid I would not see my wife and son again.

On September 1, 1967, in the headquarters one night, we received intelligence that there was a major buildup of North Vietnamese soldiers, thousands of them, coming from the North. It was to be a major counteroffensive to drive us out of Vietnam and kill everybody they came in contact with. I was in a group of seven captains, and we believed the intelligence report looked like real trouble. We were eligible for another R&R. If we took it to get out of Vietnam, we wouldn't be able to get back into the country because all the aircraft would be committed to operations. My time for my reassignment to the United States and out of the South Vietnam was September 27, 1967.

Our Vietnamese counterparts warned us don't go into the city of Hue for tea or a sandwich. We'd likely be captured or killed. I didn't even go into the village during the last 30 days. This counteroffensive was going to occur just before the Tet Offensive in January 1968.

We captains took R&R and went to Bangkok, Thailand. We watched the news, following all the news reports. Nothing had happened, but the reports indicated a large force was still coming down the Ho Chi Minh trail from Cambodia, Laos. The enemy was building large stockpiles of weapons and ammo. At the end of seven days nothing out of the ordinary was happening in Vietnam, so we had to go back. I thought this was bad news for me.

At the headquarters compound in Hue, there was a small officers club as well as an enlisted club and an NCO club. My strategy was to get to know the pilots because when it came time for me to rotate out of Hue, they would put me on a small army aircraft, about a 10-seater and fly me back to Saigon, where I would be out-processed.

I spoke to the pilots, "Look, guys, I've been out in the field. When you file a flight plan to Saigon, file your flight plan to carry us over the South China Sea, and not over the Ho Chi Minh trail or jungle."

On land, the antiaircraft guns could shoot down the plane. I didn't want to have served 11½ to 12 months over here and be shot down before I got to Saigon's Tan Son Nhat air base.

The pilots said, "Okay, Bryant, we will do our best. We can't make you any promises."

"Bartender," I said, "they can't pay for any of their drinks. I will pay for theirs anytime they visit this club." The pilots told me they'd do their best.

Shortly before I was scheduled to go home, my orders came in assigning me to Fort Polk, Louisiana. I thought, *I am absolutely not going to Louisiana.* We were coming out of the days of segregation slowly. There was nothing to do for Black families in the town around Fort Polk. I had been fighting in these jungles and I wanted to be someplace where I could spend some time with my family. I penned a letter asking the Army to please assign me some place else. They wrote me back that I don't tell them where to assign me. I wrote back if they assigned me to Louisiana, I was going to resign my commission. I didn't want to get out of the Army because I loved the Army. I didn't say anything about the racial situation there, but the Army knew what I was driving at.

A letter came back. It stated: "For cogent military reasons we are changing your orders and we're assigning you to attend the Infantry Officer Advanced course at Fort Benning, Georgia, starting January 3, 1968." Hallelujah!

September 27, 1967: my expected date of departure. On the aircraft, I knew the pilot. I had asked him to give me a signal whether we'd go out over South China Sea and fly over the water en route to Saigon under the radar until time to be detected by air traffic control.

On the plane, my heart was pounding fast. Please don't let it be a flight plan over the jungle. I was one of six on that plane.

The pilot called out, "Bryant" and gave me a thumbs up. I cheered! We flew about 200 feet over the water, well below the radar. No way the enemy could see us. The pilot raised the aircraft just outside of Saigon and we were picked up on radar. I thanked God I had made it.

I went through the briefing. I had my duffel bag, and I was sitting there in my jungle fatigues. I had the sleeves rolled up, but I felt naked. I had turned in my rifle.

Rockets and mortar fire crashed onto the airfield. The siren alarm started. I kept my seat because I could tell by the sound of it that it was not close to us. The guys around me, who had just been assigned and had never heard shots fired, dove under the benches they were so afraid. I said to them "I don't know why you guys are doing that. It's all the way down on the other end of the airfield." I sat in my seat, thinking they were not going to get me.

The big Freedom Bird, the transport, arrived. I jumped up and got in line. It was very touchy; the attacks continued at the other end of Tan Son Nhut Airfield, which was a huge air base. On that plane, I turned around and saluted. I announced, "I'm never coming back here again." We lifted off and you could hear our cheer, "Yeah!" all the way to New Jersey.

Heading back to the States, we were scheduled to refuel in the Philippines at Clark Air Force base. The pilot circled around so we could see where we were going to land. At about 25,000 feet, all I could see was a speck like a rock down in the water of this huge ocean. On that rock was a 10,000-foot runway. The pilot set it down so smoothly, we cheered again. In the airport we had two hours to explore in the terminal. It was the first time I had seen American dollars since Hawaii. I wanted to buy candy using American dollars.

We lifted off. About two hours into the flight, the pilot came over the intercom and announced an engine problem forced us to divert to Anchorage, Alaska. I thought, *Oh no!* Emily and all my family would be waiting for us. The pilot told us not to worry, he'd send a message back to the States to alert our families.

We flew into Anchorage and were on the ground for four hours. I wondered if they'd put us on another aircraft or keep us on that plane with a faulty engine? We fought for a year and this engine's going to give out on us and crash.

As one of the senior captains, I talked to the pilot. He instructed me to tell people to settle down, we'd be okay. The Air Force wasn't going to take any chances. I got on the microphone and said, "Attention, men, the aircraft has been repaired. In about three or four hours, we will be landing at McGuire Air Force Base in New Jersey.

In New Jersey, I kissed the ground and said, "I'm not going back to Vietnam again." Emily was there with my son, and I cried. My mother, my brother and several other family were waiting for me.

We took 30 days of leave, from September 27 to October 27. I snowbirded or did temporary duty, until January 3 when my Infantry Officer Advanced Class would start. The Headquarters Department of the Army required people in our category, especially guys coming out of combat, to take the course, which lasted from January to September.

One night I went to officers club to cash a check. There was one of my classmates, drunk as a skunk. He greeted me.

I asked him what was he doing. "Don't you know we have a big exam tomorrow. A four-hour exam?"

"Who studies for the exam? We've got all the answers." I reminded him they have three versions of the exam. I couldn't believe cheating was happening again. He claimed they had all three versions of it. Did I want the answers? I declined. I didn't want anything to do with that.

My brother who had gone before me on active duty, warned me not to take any of that information from guys like that because somebody will rat you out later. Instead, I had studied, studied, studied. I scored 90 percent on that exam. Some of the guys got 96 to 98. They never made a perfect score to keep suspicion down.

The advanced course prepared me for a position as a staff officer or commander in a company, brigade or battalion.

Emily had an opportunity to go back to school to become a dietitian. She would have to leave me before I graduated from the advanced course. We had a major decision to make. Both Emily and I knew the Army might send me back to Vietnam. If I had to go back to Vietnam, I wanted her to be prepared to support herself and our son in the event I didn't make it back. They were killing officers, fine young soldiers, everybody, left and right. We decided she should. She made the appropriate contacts with the program sponsor at Tuskegee Institute, now Tuskegee University, in Alabama, for a year-long dietetic internship program. It was highly selective, choosing only about 12 students to participate in this program. Fortunately, Emily was selected. I actually wasn't surprised when she got her acceptance call.

In early August Emily had to report to school and left me at Fort Benning. It was only about an hour's drive to Tuskegee, but we had to get somebody to keep young Wil, who was three years old. Emily's sisters helped us out. When one of Emily's sisters left, another sister would come. Ernestine, then Henrietta, came up and spent two weeks taking care of Wil. Her sisters cooked and had dinner on the table when I got home. I would play with Wil and eat with them. Then I'd go into my room and study because that's what the advanced course required.

Some officers were being kicked out of the course because they were not doing well academically. If they failed three or four exams, they were booted out of the Army. My two brothers had set a blazing trail for me, so I studied every night. When it came time for me to graduate from the advanced course, Emily returned for a very nice ceremony.

My big question was always what kind of orders was I going to get next. The Department of the Army representative came to talk to all the infantry officers. All of us had been in Vietnam and we were told that chances were we'd be sent back to Vietnam. The next day we'd get our assignments.

My heart was pounding—again. This time I was assigned temporarily to Fort Dix, New Jersey. The Army knew I was going back to Vietnam, but I believe they were trying to figure out which unit. I was a captain. I was qualified to serve in brigades or battalions. I earned my combat infantryman badge, so I waited it out at Fort Dix.

Emily didn't have a house or anyone to care for our son at Tuskegee so I drove Wil Jr. to her parents in Savannah. It really broke my heart to leave my son there.

At Fort Dix. I was assigned to G3 training. I knew I had to go overseas. I told Emily I wanted her to be prepared and get her certification. It required her to take a national examination.

At Fort Dix, the colonel called us seven captains to report for our assignments. Only one of us had been selected for promotion to major. Some had been selected to leave the Army for less than average performance. I was the second senior captain in terms of date of rank, but I was the only one who had graduated from the infantry officers advanced course.

Where were they going to assign me, back to Vietnam or someplace else? So many of our American boys were being killed.

The colonel said, "Captain Bryant, you have been selected for promotion, which will happen in the coming months. You will be in charge of this unit. Get these men organized. Get to work."

On that day, I was so happy to be selected for promotion. I never thought I was going to be a major. I would wear the "scrambled eggs" on my cap, which meant I was a field-grade officer which was big time in the Army. I called Emily with the exciting news.

Still, where was I going to be assigned? The next week I got a call to report to the G-1, the personnel officer; my orders were in. I felt shaky all over. The next day, the personnel officer handed me my orders. I did not want to go back to Vietnam. If I was reassigned there, I was going to resign my commission because I knew I would come back from Vietnam in a body bag.

The first thing I checked was the APO number and it read APO San Francisco 96224. I asked the colonel, where is that? Vietnam is APO New York. He told me it was South Korea. I was going to serve in the Second Infantry Division along the DMZ. I wouldn't be shot at every day like in Vietnam, but it was in the combat zone, earning me combat pay. I thought, *Thank you God, Thank you God.* I was leaving Emily and Wil again, but at least it wouldn't getting shot at every day like in Vietnam.

Chapter Eight

Korean Year of Losses and Fixing Problems

In April 1969 I was scheduled to fly from Fort McGuire Air Force Base, New Jersey, to Seattle-Tacoma International airport and go on to Fort Lewis, Washington, where I was processed for my Korea assignment to the Second Infantry Division. I was a new captain-promotable, which meant I would have one or two jobs. I said goodbye to Emily and Wil.

At Fort Lewis, I learned I was going to a light infantry battalion, the 2/38th Infantry, deployed on the demilitarized zone in South Korea in the combat zone. Many people don't realize that the Korean war has not ended. It is just a cease-fire. Near the Military Demarcation Line, we were always ready, locked and loaded. Over a big loud speaker, the North Koreans blasted propaganda daily. On many occasions we had to repel North Korean intrusion attempts, weapon fire or Claymore mines. Our job was to keep them out of the DMZ in our sector. I had a .45 revolver and an M16 rifle. I always carried a couple of hand grenades on my belt. I was assigned as the operations officer S-3, which meant I was the person to do the planning with a small team for the training and deployment along the DMZ. We would stay up there five to six weeks at a time to defend a sector. We would return south of Imjin River for another four weeks to train. We would often get new equipment, ammunition, bullets, whatever was needed.

Two Encounters with My Father

Earlier that year my mother had called and said my father had learned I was married and he had a grandson. He wanted to meet my wife and my son. I told my mother I didn't want to see him because he had done nothing for the seven of us. I wanted nothing to do with him, but my mother insisted; I never disobeyed my mother no matter how old I was.

In March, we drove to Miami. My father was delighted to see his young grandson. My son was as rambunctious as he could be. My father would pick Wil up and my son would slide right out of his lap. He took us to lunch at Morrison's Cafeteria, a buffet-style restaurant. That lunch was about the only thing he did for me, other than allow me to use his car to take my date to the senior prom.

Just before we were departing, he said, "Look, Sport," that's what he called me, "I want to talk to you in private." He asked me to walk outside with him. He said, "I'm sorry for the way I treated you boys. I didn't really provide for you like I should have. I'm so proud of you. Your two brothers Alvin and Willie and now you have graduated college, got married, and are moving up in the military. I want to give you some land."

"No thank you, sir." I had a lot of rage inside me, but I wouldn't express it to him.

"As you know I have land, being a tomato farmer, and I have several apartments I collect rent from," he continued. "I want to share with you."

"No, but thank you very much. Emily and I are doing very well. Maybe sometime in the future as my son grows up, if you're still around, maybe you could help him."

I refused to accept anything from him. We spent about four hours with him and then we took off.

The Funeral

In August 1969, I was out in the DMZ looking over toward the North when a call came on my frequency modulated radio on my Jeep. I had three radios, including a secure radio that I could talk in plain language to headquarters, so they didn't have to decode everything I said. The call came from the American Red Cross in Miami on one of my radios. A lady told me she had bad news. My father had died of a massive heart attack. If I wanted to return home for his funeral, the Red Cross would fly me free of charge from Kimpo Air Force Base in Seoul, South Korea to Seattle-Tacoma International Airport.

I thought for a moment and told her I would like to do that. I don't know if I loved my father, but I never had a chance to tell him that nor did he ever say that to me. I figured this was the way that I could bring closure.

I had been promoted to major while I was in Korea. This was only my second promotion that Emily wasn't with me. I was going from a company-grade officer to a field grade officer status, a big deal in the Army. I needed a new cap with the braids on the visor. We called the insignia "scrambled eggs." I asked Emily to buy a cap for me at the nearest military base. I had brought my Army green uniform with me to Korea.

We flew to Miami to attend the funeral. My oldest brother Alvin, my second brother Willie, and I went to the church, the same one where I was baptized at age 11. This was the first time that the Bryant brothers had returned to our hometown of Goulds in many, many years. I felt proud in my Army uniform with my new major's leaves on.

Our arrival caused a major disruption at the funeral service. We weren't too late, but they were getting ready to start.

One of the older deacons recognized us. We went to the left side of the church; my father's family was on the right side. We didn't feel we were part of the family because my father had disowned us. That day we found out that our father had two other families in addition to us. He was a wealthy Black man with his land, his apartments and his job as a foreman at H.L. Cox and Sons Potatoes.

The deacon greeted us saying how good it was to see us boys. He asked, "What you're doing over here on this side?" We told him we'd sit in one of the other regular sections. He replied, "No, no, no. You are John Henry Ashley's boys. I want all of you and your wives to come on over and we will make room for you to sit with the family."

My oldest brother said, "No sir, we'll sit over here." I was the third brother. I didn't do any talking.

"No, no, no," the deacon repeated. To the family, he said, "You all slide down for these boys."

We sat there at the service, went to the cemetery, and buried him. I didn't know all the politics between my oldest brother and my second brother and my father. I didn't want to know. At least I had a chance to say goodbye to my father, but it really hurt me. It was quite a tough time. After the funeral, I took Emily back to Alabama to continue her program.

Life in the DMZ

I flew to Seattle and on to Seoul. When I arrived, I didn't realize that since I had been promoted to major, I wasn't going back to my 2/38th Infantry unit. There was a tremendous need for field-grade officers of my rank, especially those who had served in combat in Vietnam. The big question was where would the division assign me, this Army major who had fought in Vietnam and was airborne trained.

The orders came down. I was to be reassigned to the 2nd battalion, 9th Infantry (Mechanized), North of the Imjin River as executive officer. The battalion was permanently stationed along the DMZ. The difference from the previous battalion was I wouldn't rotate out for five weeks but would stay in this battalion for six solid months in the combat zone, confronting the North Koreans. We defended a wide area, several thousand feet across the front in our sector. We positioned soldiers in foxholes every three feet all the way across the sector.

The Battalion Commander

At Camp Weitzel all we did was work, eat, and sleep, 14 to 16 hours a day. Of course, I learned every job in the battalion. As the executive officer, I was the person held responsible if something went wrong in any areas of the unit. When I reported in, the battalion commander, who was a lieutenant colonel, welcomed me, informing me I was second in command.

Along the DMZ, we had guard posts that were over 15 feet high so we could look over into North Korea. With a D-Scope, I could look right in the North Koreans' eyes.

We had to follow the battalion policy that one of us, the commander, the executive officer, or the operations officer, had to be on the DMZ every night to inspect the deployed troops, checking every soldier in the foxholes once before midnight and once after midnight. We were going to have some long days.

Almost every third night, when it was time for the commander to be on the line, he'd ask me to go to the zone for him. He was my boss, and I couldn't say no. I didn't have anything to do anyway because all we did was work, eat, and sleep. I answered with a "yes, sir." After six or seven times asking me to fill in for him, I realized he never intended to be on the DMZ.

One night we were with an inspection team led by a brigade representative. We had to climb vertical steps about 15 to 20 feet to get into the tower.

My colonel went first, and I started climbing two or three steps behind him. All of a sudden, he lost his grip, came tumbling down and knocked me down to the ground.

The brigade representative, a field-grade officer, said, "I smell alcohol on him."

I quickly said, "Let's cancel this portion. I'll go up and check. Sir, you take the brigade representative back."

The brigade representative must have conveyed that information to the brigade commander immediately. The very next day, I came in around noon, my 30-minute window for lunch.

I was sitting in the officers club with my food when a brigade officer came over and said, "Major Bryant, the brigade commander would like to see you."

I said, "I beg your pardon?"

"You heard me. He wants to see you right away."

My heart started racing. I asked, "What is the nature of this meeting?" He didn't tell me anything. I was afraid—what did I do? Were they going to relieve me? This had to be something bad by the way he was summoning me to come to headquarters.

At headquarters, I saluted and was told to stand at ease. First the brigade commander commended me for my work as XO. He continued by saying he had bad news. He was going to relieve my battalion commander of his duties within one hour.

I said, "What?"

"The assistant division commander is en route via helicopter. Here's what I want you to do. First of all, everything that's happening now—keep it secret. Next I want you to get the colonel's Jeep and his driver. Put the top on the Jeep. Go into the hooch you share with him and pack all his personal belongings and put them on his quarter-ton trailer, tie them down, and cover it up. By the time the assistant division commander gets here, I want his Jeep pulled up in front of the headquarters. Then I am going to call headquarters, and when the general arrives, he is going to relieve your boss of command. This is highly classified. Take off. Get it done."

I hurried, found my driver, and briefed him in confidence. My driver and the battalion's driver got everything packed.

Here came a UH-1 helicopter, the kind with the glass bubble. I could hear the whirring chop approaching. They were only two seats, for the pilot and one passenger.

I brought my battalion commander to headquarters. The brigade commander told me to wait in the other room. He insisted on the highest degree of secrecy. He named me acting commander for a few hours until another battalion commander arrived to replace him. *Oh, my goodness. If they are doing this to this to a lieutenant colonel, they would stomp me if I did something like that.*

So sure enough, they had cleared all the personnel from the front of headquarters; they relieved him of command, put him in his Jeep with the top on so no one could see in, and then they shut the door. They flew him to division headquarters in the UH-1.

I found the command sergeant major and briefed him I was the interim commander for a few hours. The same helicopter that took my boss came back with a new battalion commander.

This incident really got my attention. I was doing the right thing. I didn't drink because of my mother's rules, which carried over to me in the military. When I went to the officers club, I'd have a ginger ale or a glass of coke and tell them to put a cherry or a slice of lime to give the appearance that I was socializing. They had happy hour every Friday night because there was nothing else to do.

Others would drink a lot, but not me. We had one officer get inebriated who went out and sat on the wall around the club. He leaned over, fell, and rolled down the hill, breaking both ankles. We had to get that squared away ASAP and stop the excessive drinking.

Parts for the Motor Pool

My new battalion commander brought in numerous policy changes. I was also to be his chief maintenance officer, even though we had a lieutenant and warrant officer officially assigned to those positions. I was the guy to keep those armored personnel carriers, M-114s and the wheeled vehicles operational. The commander wanted to keep all vehicles operationally ready. Under Division policy, we weren't allowed to have more than six vehicles down, or on dead-line, in Army terminology. We had over 100 vehicles in this battalion and could only have six down. He reminded me it wasn't his policy, but the division's policy. He recommended I meet with the maintenance warrant officer and work out a plan to meet that goal.

The biggest problem was not getting the vehicles operational. The trouble was keeping spare parts in the battalion. We would get them and for some reason, they'd disappear. I asked my maintenance warrant officer, CW-3 Liam Kokason, to find out what was happening to our spare parts.

Kokason was a Black guy from Australia with an accent. I talked with him and his immediate superior, the battalion motor officer, who was a young first lieutenant. I knew the lieutenant didn't know anything about these spare parts. Kokason had been around for close to 20 years. We discussed division policy and how we needed to stop the loss of parts to meet policy. He asked for a few days to investigate. He found the Koreans ran a trash truck throughout the battalion every day. They entered around 9:30 a.m. and they were coming out of the compound between 11:30 and 11:45 a.m. Kokason had a hot tip that's where the spare parts were going. Also the military police did not inspect the truck when it left the post.

When I let the battalion commander know, he brought in the military police, the criminal investigation division or CID. We also involved the South Korea military police. I set a date and had all the vehicles of military police hidden around the post. When the trash collectors worked their way up to the front gate, the military police vehicles drove behind the trash truck and blocked it. We put a big tarp down and pulled all the trash out. A large array of our new spare parts was stashed in the trash. The military police made the arrest and busted that ring right there. Thanks to Chief Warrant Officer Kokason, the battalion commander was very happy.

Moral Consequences

The only place we had for enlisted men for relaxation along the DMZ was Recreation Center #3 with steam rooms, ping-pong tables and card games. Donut Dollies ran the Center in the evenings. However, soldiers also traditionally took two-day passes to socialize with the locals down in the villages south of the Imjin river. Officers were not authorized to do that. When the soldiers went south of the Imjin River, they had to leave their weapons up north of the river. The division had this crazy policy: No more than six soldiers in any battalion could contract sexually transmitted diseases. My new commander handed this problem to me.

I said to my commander, "Wait a minute, sir, are you assigning me as the morals officer in this battalion?"

"That's about it, Major," he said.

"Sir, that's a tall order. I have no control over those guys when they go down in the village." He didn't care how I did it, but he was holding me responsible for keeping the number down to five or less. I had to pull together a strategy quickly.

I told the command sergeant major of the battalion, the highest ranking NCO in the battalion, I was going to suspend all passes for every soldier for two weeks. No one would take leave or get a pass, unless it was an emergency and I personally signed off on it. I announced a troop information program about sexually transmitted diseases to be conducted over the next two weeks. The command sergeant major and the battalion surgeon created a curriculum with instructional aids to teach all soldiers about these serious diseases. Unfortunately, some of them can take a soldier's life. Soldiers can't deploy back to the United States with these diseases.

The moment I put that suspension in place, you could hear soldiers screaming from Korea to Los Angeles. Some even went to the inspector general of the division about my plan. They argued they had female friends down in the village. I told them I didn't care. No leaves or passes!

Next, every soldier was issued at least three condoms before they went down to the village in addition to the classroom instruction. We had a headquarters company, and three line companies, A, B, C, and a combat support company deployed along the line. I told our headquarters company personnel, if any soldier contracts the disease, he's going to be deployed along the line in the DMZ and I'd put him in a foxhole. No one wanted to do that because the temperature in Korea on a nice cold night was -20° degrees with a wind chill of 28 to 30 below zero. Even with our cold-weather gear, Mickey Mouse boots and field trousers over wool trousers, parkas, masks, and Arctic mittens, if you weren't moving, you would get cold in a hurry.

My headquarters soldiers worked in the mess facility and the motor pool. They lived in a hooch with heat in their rooms. They'd have to give that up if they contracted a disease and go serve on the DMZ.

My driver, who I thought was the best driver in the battalion, was a young man from Arkansas, a Specialist-4. I'll never forget his freckles and red hair. He could really maintain a vehicle. He had asked for permission to get a heater out of a five-ton wrecker and put it in my quarter-ton Jeep to keep me warm. He was a doggone good driver. I never had a problem with my Jeep when he was my driver.

The young man was highly capable and polite. I promoted him from PFC to Spec-4 after about a month. Early one morning, I was sitting in my office about 5:30 a.m. He usually picked me up at 6am to take me to the officers club for breakfast.

At 5:40, he knocked, came in, and saluted me. "Sir, I have a problem. Remember the briefing that you gave all the men about going south of the river and getting in trouble? You said you'd send any headquarters company person to serve along the DMZ and the front if they contracted VD."

"It is a very active policy."

"Sir," he said, "I went to the village, and I have a STD."

I said, "You did what!"

Mind you, I had already sent about eight soldiers from headquarters to the zone. The battalion commander was aware of the policy and said it was a tough policy, but as long as I keep the STD numbers down, it was fine with him.

"Please don't send me to the zone." He started crying.

It really broke my heart. I asked him, "Why didn't you do what I told you to do?"

"Sir, to be very honest with you, the young lady told me that I was her first and I didn't have to worry about anything like that. I didn't listen to what you said or what was said during all that instruction during those two weeks. I took a chance." Crying, he said, "Please don't send me on that line. It's cold out there."

"Stop crying," I said. "Go back to your barracks. Do not talk to anybody else. Who did you tell this to?"

He said he hadn't told anybody.

I went to see the battalion commander to explain what had happened to my driver, who, by the way, also took care of his Jeep. He had wanted to take my driver, but I had refused to give him up. I told him I was going to make a decision regarding my driver unless he overrode me. I wasn't going to send him to one of the foxholes.

"Pull up, Wil, pull up," my commander said. "You want to violate your own written policy?"

"Colonel, this is an outstanding young man. He is only 18 and a half years old and he has never made a mistake. He told me that he was a virgin when he went down to the village, an honest mistake. Sir, I believe this young man. I'm going to make an exception."

"Wil, what if someone else found out about him contracting the disease and your making an exception for him? You have already sent our other headquarters soldiers to the DMZ."

At that moment the only people who were aware of what had happened were my driver, my commander, the battalion surgeon and myself. I knew it wasn't right, but this was an outstanding young man and soldier. I argued I didn't know how long he would stay in the military, but whatever he does in life, he was going to be successful.

He was a hard worker, he studied hard and was one of the best soldiers in this battalion."

The battalion commander stopped me. "You don't have to tell me that. I know it, but it is your written policy."

I replied, "I will take that chance. If it happens, you hold me directly responsible." He said he'd do that.

I called the battalion surgeon and my driver. I explained what was going on. I told the surgeon I wanted him personally to treat him, get him healthy, but to zip it up or don't say anything to anybody.

The surgeon knew I was in charge of his officer efficiency report and I knew he wouldn't say anything to anybody. He also knew what kind of young man my driver was. He said he'd take care of my young driver and get him back to duty as soon as possible. He said the new medication would speed up the process so I put my driver on quarters for at least two days and gave him some unassigned duties up at headquarters, non-vehicle tasks.

It never happened again. I told my driver I was making an exception to my rule for him. I told him always think before you act. He promised it would never happen again. We shook hands on that. We never had that problem again.

The Orphanage

One additional duty of mine was non-military. The U.S. Army had helped to build orphanages throughout Korea. We adopted an orphanage in Korea south of the Imjin river, outside the weapon zone. There were about 30 young girls from 6 to 12 years of age, children of South Korean women who had given birth from their association with American soldiers, Black or White. It was taboo in Korea, so they abandoned the mixed-race children. The battalion commander assigned me the responsibility for support of that orphanage. I visited the headmistress, and I saw all of these children with no mother or father, no relatives. Because of my fatherless background, I decided to emphasize on an existing policy, and to ask for each officer to donate voluntarily to a special fund to be presented to the orphanage each month. We bought books, clothes, and toys—everything we could to enhance their well-being. I had an officer assigned to do my legwork for me, which required quite a bit of coordination.

As in Vietnam, in Korea I didn't need very much money. Mine all went back to Emily or the Savings Deposit Program, which earned good interest. If I kept $75, I would put $50 into the pot of money for the orphanage. The battalion was invited to visit the orphanage once a month to be with the kids. I'd take the funds down to the headmistress, the lady who ran the school. The kids would put on a show for us. Their smiles and happiness warmed my heart.

Searching for North Koreans

At 5 p.m. every day all indigenous personnel had to be off our compound and south of the Imjin River. Once there was a sighting of supposedly 8 to 10 North Koreans on a rubber dinghy in our sector. The assistant division commander needed a combat-trained officer to lead a reinforced rifle platoon into the suspected area to find those North Koreans.

When my battalion commander said I was the only combat-trained officer in the battalion, I reminded him I was a field-grade officer, a major. A captain or a first lieutenant could do it, but unfortunately I was the only one with a combat infantryman's badge.

Landmines had been planted along the DMZ during the Korean War. They tried to clean that area up, but they didn't get them all. Landmines and snipers were my greatest fear, as they were in Vietnam. If I had to be in a shooting operation, I demanded 10 to 15 mine detectors.

This was a major operation. We had a fixed-wing aircraft, a spotter, circling close by like we had in Vietnam. If we saw an enemy target, I'd call the spotter with the enemy's coordinates. He would have the F-105 Air Force fighter jets, Thunder Chiefs, hit them with bombs, so we could go in and get them.

We searched that area for seven or eight hours, all day. The circling aircraft didn't see any movement. They had done what they were going to do—we saw tracks—but they had gotten out quickly. Nobody stepped on a mine or got hurt, so it was a successful operation.

Moving On

My time in Korea slowly came to an end. I found out that I would be returning to the United States the end of May 1970. The next question was where was I going from Korea. I received word I was going to be assigned to ROTC duty at Howard University in Washington D.C.

Wow! I wondered how long it would be. I guessed probably two years and then I'd move to another combat zone. The assignment thrilled me because I always wanted to go back to school for my master's degree and this was a great opportunity to do that. Both my older brothers also studied there. My second brother had graduated from Howard University Dental School in 1968. My oldest brother was there in medical school in his last year. He was president of his class and would be graduating in June, 1970.

I reported to Howard University on July 1, 1970. I was able to attend Alvin's graduation and sign in to the Department of Military Science at Howard University. I was assigned to be commandant of cadets. The professor of military science told me I had a very distinguished background, also complimenting my public speaking ability, so my skills continued to move me forward.

Chapter Nine

Commandant of Cadets Meets the Black Panthers

For the Howard University position, we found a neighborhood with a school young Wil could walk to and a big playground close by. Our new townhouse was in Forestville, Maryland, about 45 to 50 minutes from Howard University and close to Andrews Air Force Base, Maryland.

At Howard, I hoped for a high-quality staff. The professor of military science, Lt. Col. Maurice Williams, had been there for several years. The executive officer was Maj. Jimmy Williams. Maj. Chuck Lawson, S-1, personnel officer, was soon to be replaced by Capt. Willie Gore. All of these officers became good friends. The professor of military science decided, based on my experience and my background, I would teach the junior cadets. As commandant of cadets, I was responsible for all their preparation in the advanced course, the third year. They were paid ROTC students so they would become new second lieutenants like I had been during my college days. That position was quite an honor.

The senior noncommissioned officer was Sgt. Maj. Ralph Odom, an airborne-ranger-qualified soldier. Shirley Adams was the department secretary, succeeded by Linda Blakely. We were assigned a new S-4 logistics officer, Capt. Marcus Woods, who was a graduate of Morgan State University.

Howard was a different—civilian—environment for me, a gung-ho soldier coming out of Korea's DMZ weapon zone. My approach to accomplish tasks wouldn't work on the campus. When I gave an order in the Army, it was carried out right away. At Howard, it was 180 degrees in the opposite direction, so I had to learn the system. For example, to get an invoice approved sometimes took three or four weeks. I managed the budget for the Department of Military Science, more complex because I had an Army budget and a Howard University budget. To get money from the university system was like pulling teeth from an elephant, until I found out who approved the invoices. When I established a rapport with her, I was able to get our invoices through. We needed to give awards and plaques to students and for our activities. Our goal was to recruit new students for the ROTC Program, to build it to a respectable level.

At Howard in 1970, the Vietnam War was very unpopular. Campus protests raged. When I reported for duty, Lt. Col. Williams advised me not to wear my Army uniform or maybe only one day a week. I was proud of my Army green uniform and wanted to wear it. Teaching the juniors I wanted to set an example for them. I was also motivating students who were freshmen and sophomores who were thinking about going into the advanced program as well.

"Bryant," he said, "if you wear your uniform and there is an incident on the campus, that could be a problem and ultimately a reflection on your judgment."

"Sir, I understand you. Unless you give me a written directive not to wear my uniform, I'll wear it. I'll deal with the protesters."

"You know I can't give you a directive. I have no justification, no legal standing, to order you to wear civilian clothes," he said. I answered I was a big boy and I could take care of myself.

My very first week on the campus I walked over to Cooke Hall to get my lunch in one of the cafeterias in the men's dormitories. I was in uniform with my 2nd Infantry Indian Head patch on my right shoulder and my Vietnam combat infantry badge on my left chest. A group of Black Panthers, who were very active on campus, sat blocking the entrance.

"Excuse me, gentlemen, I'd like to get through," I said. These guys started screaming all kinds of nasty things at me. I said, "Wait just a minute, gentlemen, I don't know any of you. This is the United States of America and I wear this uniform to do my part to protect the democracy that you all enjoy and that allows you to sit down and do whatever you want. I'm gonna walk through here. If one of you touches me, then there is going to be a problem." I adjusted the bottom buttons on my Army green uniform to give the illusion I had something under my jacket.

One Black Panther said, "Oh, this guy, he probably has a gun. He's probably gonna shoot us."

"Move out of the way," I said. "Don't touch me." They decided to let me go. I only said, "Carry on, men." They sat there for about two weeks.

Every day, they'd say the same thing, "Here he comes, ladies and gentlemen. Here's this warrior. Be careful. He may have a gun."

Every day, I said, "Good day, gentlemen, have a good day." They never touched me.

Finally, there was always student unrest on the Howard University campus, not only Vietnam war protests. As I recall, there was a protest or two where students took over the "A" (Administration) building. Further, during exam week, there were always bomb threats called in to campus police; after the first time dealing with this situation, I always administered my exams prior to the week designated for exams. Among all of the HBCUs I visited over the years, Howard University was, by far, the most politically active.

One day I saw a friend of mine, a classmate from Florida A&M University, who was an FBI agent, walking around campus in a suit with a white shirt and a nice tie. He'd been a tight end on the football team, and he was a big guy, over 6 foot.

"Hey, Big Ray," I said. "What are you doing on campus?" He told me to shush because he was working undercover. I almost laughed. "Undercover! Everybody knows you're an FBI agent in that suit. Put on an old dirty field jacket and don't shave for six or seven days and then you'll blend in." He told me he'd take that under consideration.

During my ROTC assignment, I told Emily I viewed our Howard University assignment as an opportunity for both of us to get our master's degrees. All I'd have to pay was my registration fees because I received remission of tuition as a faculty member. Emily would have to pay, but that was no problem—I could take care of that. I also had my VA benefits, which gave me a check every month to supplement my military pay.

My second oldest brother Willie, who had just graduated from the dental school, told me about a new master's program, student personnel administration, on the campus. He suggested I go see

Dr. Carl Anderson, vice president for student affairs, who was a fraternity brother of ours. Looking forward to when I left the military, I wanted to work on a university campus as a president or vice president, dealing with students, so this master's would be an ideal program.

Dr. Anderson gave me the brochure for the new program, which was going to kick off in January, 1971. About six months after I arrived, I started in the master's program in student personnel administration.

Howard also offered Emily a great opportunity to get her master's. Again I didn't know what the Army would do with me after Howard. I wanted her to be prepared educationally, if I was sent back to Vietnam or deployed in a combat area. If I became incapacitated, I wanted her to be able to take care of the family. I got her registered and all squared away in her department. All she had to do was to report for classes that next semester to start in a Master of Science program, leading to a master's degree in nutrition. I don't know how Emily did it. She was fully employed at Hadley Hospital as a brand-new dietitian and began her graduate program while caring for our home and Wil Jr, and then giving birth to our daughter Lisa.

I was very proud of Emily working at Hadley Hospital because she was doing a great job dealing with folks there. She, of course, had her challenges. She held her staff to high standards of cleanliness and strict nutrition guidelines.

Young Wil, who was six years old, began his athletic career. He was a good student, a good boy. We tried to provide him with activities to become a complete student. He took after Emily; he was a happy, pleasant young man. He didn't fuss or get into fights. Emily was the one who introduced him to football. I give her all the credit for getting him started in his football career.

Lisa was born at Andrews Air Force Base Hospital in March, 1972. It was a happy day in our life. Our two children would see us at the table, doing our homework while in graduate school and young Wil would sit on the floor doing his work for school. We demanded he complete his homework first before going out to play. We were trying to set an example for him. Lisa was, of course, only a baby, but she was soaking it all up.

In 1970, we bought Emily a white Volkswagen bug to drive to work. I can see Emily, dressed in her white dietitian uniform with her white stockings, putting Lisa in her car seat and taking her to the babysitter.

We lived in a U-shaped complex of Forest Park townhouses, and the parents looked out for all the children. We didn't have to be too strict on Wil; no predators or lurkers were around during this time. I really didn't understand all Emily did to get the children ready each day, until she went to a conference out of town. I had to take care of Wil and Lisa for three days. I didn't know how Emily did all this every day. She would get them up, get Wil ready for school, feed them, get Lisa's hair brushed, combed and plaited. When Emily took her trip, Lisa drove me crazy—as soon as I got her hair all plaited, she'd pulled it all apart. I thought, *Oh, my goodness, Emily is going to have to come home.*

One morning I had Lisa in my arms, I had her baby bag, and I had to take her to the babysitter. There was a picture of Emily on the wall in her dietitian uniform and Lisa looked at the picture and shouted, "Mommy." Oh my, that caused a big lump in my throat.

I got on the phone and called Emily and told she had to come home because this was tough. She told me to hang in there. I think she cut her trip short by one day. After that experience, I had a new appreciation for all she did with our two children.

At the university, we received a new professor of military science, Col. Samuel D. Stroman, a very strict Army officer, and much older than his predecessor. I could tell he didn't like the way his predecessor had performed his job.

When this new professor came on board, he sensed that I was a strict disciplinarian about class work, their uniforms, and their obligations under the ROTC Army contract.

One semester I gave one student a D grade; if it stood, he'd have to leave the program and be subject to the draft. I told him if he accomplished certain requirements, I would reevaluate his papers and perhaps remove that D grade.

When I gave an assignment, I told my students, I expected it to be complete because they were being paid tax dollars from the people of America. They had signed up voluntarily to be in the ROTC program and I expected them to work hard.

The student with the D went to see the new professor to complain. The professor came out of his office into our open office with cubicles where I was right in the middle. He started yelling at me right there in public. I asked him to step out in the hall to discuss this matter. Out in the hall, he really got my dander up, but I always said "sir" first. I didn't like how he was talking to me.

"Sir," I said, "First of all, don't yell at me when you speak to me. Second, don't talk to me that way. My mother never talked to me that way." I continued and told him he should have asked for my report or my opinion why this student earned a grade of D. I explained what this student could do to repair his grade. "Sir, will all due respect, don't ever yell at me again because I don't tolerate anybody yelling at me," I said. "I want to be treated with respect, and I will treat you with respect." Maybe I was too direct with him, but I had to stand my ground.

The colonel said, "Major, I appreciate that. You go ahead with your plan and keep me apprised of the situation." I told him I'd handle this problem. If I got to the point where I couldn't, I'd get back to him. It never happened.

On Christmas morning that first semester Col. Stroman called my home at 8 a.m. He demanded I submit my grades to him, not directly to the dean of the College of Liberal Arts. He wanted them that morning. I drove in, luckily there was no traffic, and gave him my grades. I suggested he put his new policy in writing.

He told me I didn't tell him how to do his job. I closed my mouth, said, "Good morning, sir," did an about face and walked out. I just knew this colonel would rate me very low when I received my first officer efficiency report.

When the time came for the Howard University ROTC Instructor Group to nominate one officer to receive Leo Cobb Memorial Award for outstanding teaching and leadership throughout the Department of Military Science, Col. Stroman sent a note to his executive officer to nominate Maj. Wilbert Bryant. I thought, maybe I was doing the right thing.

In January 1973 President Nixon withdrew U.S. troops from Saigon, ending our involvement in the war. What did I feel? Relieved that more young men and women wouldn't die. As a military officer, I had disagreed with how the war had been conducted at the highest level. President Johnson was reported to be selecting the military targets, then advising the Department of Defense instead of listening to his military advisors. Supposedly he refused Army leadership's desires to bomb Hanoi and cut off supply lines. I was proud of my service but so glad to see the end of our servicemen and women dying there.

In my fourth year, when I was getting ready to leave Howard and received my first and only officer efficiency report from Col. Stroman, he gave me an outstanding. He made the comment that "this officer has general officer potential." I was surprised and pleased. I believed he appreciated me standing up to him. If I had allowed him to run over me, I never would have made it out of there successfully.

Our friends helped us get through those four years, the Handys and the Goodmans. Air Force Maj. Burrell Handy and his wonderful wife Madelyn had two children, a son and a daughter. Their son Darren was best friends with our son. Percy and his wife, Antoinette, Goodman was a physician who had graduated from Howard Medical School and he was a classmate of mine from Florida A&M University. Good friends helped us to continue my Army career as I adapted to an academic environment. We also had fun socially with our friends. We did get a chance to go out with our friends into downtown D.C. to Blues Alley to see the big entertainers who came to town like Harold Melvin and the Blue Notes and many Motown groups.

The other key duty, while at Howard University, was ROTC summer camp every summer for six weeks. For us to participate I had to go to Indiantown Gap, Pennsylvania, which was near Harrisburg. We went there for two summers to train cadets from all over the northeastern United States. It was really a great experience for me.

Every year we had an ROTC ball held at Walter Reed Army medical facility's officers club. As commandant of cadets, I coordinated it with a lot of help from the entire ROTC cadre. Emily and the other cadre wives were invited guests, with the menu being selected by our staff, in coordination with the officers club.

The cadets were in dress uniforms with all their insignia, braids, patches and their dates in formal evening gowns. We had a live band and played music of the times, i.e., Motown music! A few awards were presented to some of the cadets and a fun one, recognition of Miss ROTC. (During my time, we didn't have female cadets. In my last year we received our first female cadet.)

Howard University whetted my appetite for academia once I understood how it operated. I remembered again how good my college years had been. I loved watching the students come in on day one, mature, and graduate. As they had learned and grown, I felt I had made a small contribution.

When I received my Master of Education degree, I joined Kappa Delta Pi Education Society. Instead of a thesis I took an additional three credit hours in the master's program. I had to pass a three-hour competency exam at the end of my academic program to get my degree. In fact, when the exam was administered, I was the fourth student to complete it, finishing in an hour and a quarter.

About a week later, grades were posted. I knew I had done well; when I opened my exam, I had earned A+. I was proud. Little did I know that Emily had done better in her master's program than I had. She wrote a thesis, she was working and taking care of the children and she earned all A's in her graduate program.

Young Wil found out. He saw the transcript for my grades and for Emily. I earned three B's in my master's program, a 3.72 GPA. Wil told me Mom was smarter than me. Wil got into everything. I told him women are smarter than men anyway, so yes, your mother was smarter than your dad. We posed it as the challenge for him to beat us out when he reached that point in his life.

We graduated in April 1973. It was the first time both of our parents attended our graduation. My mother and Emily's mother were sitting in the stands. I was a happy individual. I worked hard at the undergraduate level because my mother had made such a huge commitment sacrifices to support me and she was there to see me get my Master of Education degree.

My Howard assignment prepared me for new challenges. I had four outstanding years at Howard University. We acquired good friends that have lasted a lifetime. I also had outstanding students graduate during that period. Joe Chesley, Harry Brooks, Herman Chesley and Damon Marshall, all went on to have successful careers in the Army. Post-Army they also did very well. I am still in touch with them. I believed I boosted them with my high standards.

Near the end of the ROTC assignment, my orders came down. I was selected to attend Command and General Staff College at Fort Leavenworth, Kansas, that prepares students for assignment at the battalion, brigade and division level. Unfortunately, I was uprooting my family again. I wanted to make sure that young Wil was prepared for that because he had made a lot of friends in Maryland. He took it in a very positive light because of our positive attitude about the new assignment.

At Fort Leavenworth the Command and General Staff College was one year of intensive study, with over a thousand students in the class of 1974–'75.

We drove from Washington D.C. to Fort Leavenworth, Kansas. We lived on post in a village named Kickapoo Village. We had a three-bedroom house. Next door was another Black family. I thought it curious the Army still had a tendency to put the Black officers together. Our next-door neighbors were Maj. Everett Rochon, quartermaster officer, and his wife Jackie, both graduates of Southern University in Baton Rouge, LA and their two children.

We had heard that Midwest beef was the best. Emily and I teamed up with Everett and Jackie to buy a side of beef. We placed our order, having it cut to our specifications. Emily took care of all of that because of her degrees and all her training in nutrition.

Our kids in the housing complex were close in age, except Lisa, who in 1974 was two years of age. She rode around the complex on her Big Wheel and had girlfriends she played with. Emily was involved with the wives in our housing complex. With her outgoing personality, she could make friends with anybody. We had a lot of parties among the families within our housing complex. Our fun and parties were all on post.

Wherever we were assigned, Emily always found a job. She signed up to do substitute teaching in the local community school, not knowing how very cold Kansas could get. There was snow on the ground, a foot and half to two feet of snow regularly but we, the kids especially, had so much fun with that.

Often we'd get a call early morning for her to teach. On some of those very cold mornings, she'd announce she wasn't going out of the house. She'd tell me to take the call and to tell them she was ill or busy. When I said, "Emily that's not telling the truth," she said, "You just do what I tell you to do."

For me, CGSC was a highly competitive school. First of all, it was headed up by a two-star general who was the commandant and had a full staff to manage those thousand-plus officers in the class. Some officers rented a trailer to study in Monday through Friday evening, away from their families. After our dinner, once I finished my activities with the children, I would go to the special room Emily had set up for me and I would study until 12:30 a.m.

At Fort Leavenworth in our student mail box, all of the officer students had to watch for a form called a Blue Goose, an assignment directing you to take the lead on the next day's lesson for 75 to 100 classmates. Not the next week—the next day. You had to be up to speed with your studies, or you couldn't handle that Blue Goose. We only had one evening to make all arrangements for any aids or a film with the NCO before we went home.

I was lucky with my first Blue Goose assignment; it was the Five Paragraph Operation Order. It was used to prepare for any unit either at the platoon, company, battalion or a brigade or larger unit operation. At Florida A&M University our instructors drilled the order into us.

I figured out what teaching aids I needed and made the appropriate contacts. When I got on the platform the next morning, I was waiting for them. I had my overhead transparencies ready to flash on the screen with my key points. I called the class to attention and directed them to take their seats.

Our professors taught us at Florida A&M "to stand and be seen, speak and be heard, sit and be appreciated," meaning when you get up, you know what you're talking about, speak with authority, and you will get your point across. My first Blue Goose was highly successful.

There was not sufficient parking in our complex, so we formed carpools from our homes to headquarters. One day I was in the car with a veterinarian and three other officers. On this day somebody's dog darted out in the street. By the time I could get my foot on the brake, I had clipped the dog's hip. I stopped the car right away. I felt so bad. The vet said we needed to get this dog to the base's vet. I told the other guys to walk home. I got home and cried. It hurt me to my heart. We had dogs when I was a child. Thank goodness, the base vet did surgery right away. Emily gave me a lot of grief—"You've got the vet in the car and you hit a dog!"

While I was in that intense school, the children, especially young Wil, had so many activities and sports. Wil continued to be a good student. He earned all A's and B's, which was our criteria for him to participate in sports. Lisa would be at Wil's football games in her little snowsuit. She would mimic the cheerleaders moves on the sidelines. I knew then she would end up a cheerleader, too.

Emily's best friend, Betty Wright Blakely, came out to visit us when we had a foot and a half of snow on the ground. That night we had a party within our court. The ladies were dressed up fancy in new outfits. Emily was in a brand-new yellow wool outfit, but as they stepped out—boom, they slipped on ice and hit the ground. Fortunately, the snow cushioned their fall. The moment was funny, but it was not funny. I helped them get up, cleaned the snow off and we went to the party. Everyone had a good laugh over Georgia/Florida girls in the snow.

That year went by fast. Because of the intensity of the studies, the work and the pressure, there were over 80 officers who filed for divorce from their spouses. Also we knew that when we left there, we were going back to the real world of the Army. We knew we would have to perform at the battalion, brigade, division and corp levels, but the school prepared us for this next step in our career.

I had done well, but the top student in the course was a young Army captain from West Point, Wesley Clark. He eventually became a four-star general. After his Army career, he ran for president seeking the Democratic presidential nomination. At the school, he received the Marshall Award, the top award. I met him several years ago at Fort Myers, Virgina in the barbershop and we had a good talk about Kansas and his presidential run.

Since we were getting ready to graduate, the next big issue, as always, was our orders. I received a call from the U.S. Army Military Personnel Center and was told to go to the closest air base and take a flight physical. If I passed this physical, I'd be assigned in a unit requiring me to fly. The position was in England. I thought, Wait a minute, we didn't have any Army in England? Then I remembered a course I took in the infantry advanced course, a very specialized course. I can't mention the name because of security implications. I passed the exam in this very competitive course; only 70 of the 220 students passed.

I passed the flight exam and sent the results to the Military Personnel Center. My assignment officer informed me the Army was sending me and my family to a three-year assignment. I'd be flying with the U.S. Air Force, using the specialty I acquired when I was a captain in Fort Benning, Georgia.

I was going to Royal Air Force Base, Mildenhall in England, located about 70 miles northeast of London. I couldn't believe it—a small town boy who had attained the rank of major in the U.S. Army had been selected—assigned—to England. My reporting day was mid-July. We were scheduled to fly out of McGuire Air Force Base in New Jersey.

After graduation, we were going to my mother's, but first we went to Emily's mother's in Savannah and they gave us a big send off. I'll never forget—they must have barbecued everything that was barbecue-able. We had to say goodbye to our families for three years.

I was soon to learn my mother had somehow saved the money to buy a piano each for Alvin, Willie, and me. She had wanted her boys to learn to play, but we thought it was feminine to play a piano so we didn't learn. Only my oldest brother Alvin took a few lessons. All of us had children and she wanted some of those children to learn, so I had to figure out a way to get this piano moved. It had to go in our household goods being sent to England.

We sold our big car and took Emily's Volkswagen bug over to England, where gasoline was rationed for U.S. forces stationed there. We wanted a fuel-efficient car and the Volkswagen bug was the answer.

In New Jersey somehow we missed our flight. We notified the Army. They shifted us to fly in Category Z, a commercial flight from Philadelphia International Airport to Heathrow Airport, London, England. I had to get a message to my new unit. They were scheduled to pick us up so they had to adjust to match our arrival. We went to Philadelphia and young Wil flew in the First-Class section; we were all in coach class. At Heathrow, we'd boarded a train and our sponsor was waiting us for us in Ely.

Chapter Ten

England and the VW Bug that Crossed the Alps

Landing at Heathrow, we took a taxi to the train station. I was about to give the cab driver a five-pound note, when Wil said, "You're giving him too much money."

I asked him why. He told me a British pound was worth about $2.64. My son was always curious and busy reading. Initially he, like a lot of boys, was lazy about trying to read, but we started with *Sports Illustrated,* which got him reading. I'll never forget how on the airplane in first class, an Englishman explained the exchange rate to him. In 1975 Wil at 10 years of age was as smart as he could be. I gave the driver a one-pound note.

We took a train to Ely where an Air Force van picked us up. It was a hot July day. The van pulled up to the headquarters and they opened the back so I could get out. Emily and the kids were also in the back of the van. My unit boss, USAF Col. Bob McDonald, met me. He had his "brick," a radio, in his hand. He was getting all kinds of alerts on it, and I knew this was serious business.

I was trying to be very military, talking to him, when Emily said, "Bryant, It's hot in here. We have to get off this van. Hurry up."

Oh my goodness, she's going to get me a zero on my officer efficiency report. Coming in late because we missed our flight, and now she made this comment in front of this colonel. I'm just a major. Emily didn't really have a full grasp of military rank structure and the protocol or courtesies that are given to officers of higher rank such as a colonel.

The colonel certainly noticed. "Well, it seems your family is ready to get out of the van. So why don't we take you over the officers club and get them settled in. We will get you checked in a little later." We drove directly there.

When a person is assigned overseas, a permanent change of station, another officer is usually assigned as a sponsor to help one get acclimated to the post and the cultural landscape. My assigned sponsor, Maj. Richard (Dick) Lind and his wife Linda, were a great help. We're friends to this day; unfortunately, Dick passed several years ago. Anything we needed or that I didn't fully understand, I could go to him and he would advise me how to handle the situation.

Since we were in England where folks drive on the wrong side of the road, or right-hand-drive steering, Dick gave me an English car to drive which I parked in my driveway because I was terrified to drive on those British roads.

We drove Emily's VW bug. You had to reverse your thinking and look to the right but drive on the left side of the road, not to mention traffic circles or the round-abouts. The British were not sympathetic or patient with Americans on their highways. Special license plates on our car identified us as Americans. Many times, they would honk their horns at us if we were to slow to enter the roundabout.

We stayed three weeks in temporary housing in the officers club on the base. In the Army it's called a post; in the Air Force it was called a base. In this temporary facility, there were crews on crew rest. Also there were families like us waiting for housing and people who were waiting to transition back to the United States.

One Monday when I came home from work, a little short Air Force major walked up to me and asked if I had two children. I answered I sure did. He commenced to complain that my children had been running up and down this hall, making a lot of noise in this transient quarters building and disturbing crews who needed eight hours of crew rest because they flew aircraft daily. I stopped him, saying he had the wrong party. It wasn't my family. He insisted it was my children. Then I got mad. I started raising my voice. Of course Emily heard the commotion and pulled me away from that officer.

Later, we found out there was another Black family who had two small children on the floor. That family allowed their kids to run up and down the hall. Emily never allowed our children to go out in that hall without her. I was so mad. I told him to not to ever approach me like that again. If he did, I threatened to knock him out.

There were tough times with race relations during those days, even in the military. He had assumed because I was Black and had two children that it was our children.

I think the incident probably spurred my command into getting us into housing faster. Luckily they moved us into base housing right at Mildenhall Air Force Base so I really didn't have to use my car to get to my job. I would put my flight helmet in my flight bag and walk from my house to the unit's headquarters. I'd get my briefing and report to the alert shack.

I don't like to dwell on race relations, but I was the only Black officer assigned to that unit. It didn't bother me one iota. No Black NCOs were assigned to that unit either. As always, all I wanted was the opportunity to learn my job and do it well.

All of our battle staff teams were headed by a full colonel. I had never been in a structure like that before. All these colonels worked for a senior colonel.

Emily helped me immensely. She was very active in the wives club. Here in England like our other postings, they loved her because she had that outgoing personality and was always so friendly, even with officers' wives whose husbands were lieutenant colonels or colonels.

The first year I had to get acclimated to using gas ration coupons to drive our car. We were alloted a six-month ration. Emily loved to shop and there were so many great things to purchase—English and Irish crystal, silver, and china in the officers club or from the different villages in our area. Word filtered out to the other wives she was going lots of places. On the wives' shopping trips, Emily always drove, using our gas rations.

About two months into my command, Emily informed me we are almost out of coupons. I couldn't believe it. She replied she needed some more coupons. I had to ask the big boss, the colonel in charge of our unit, to get a supplemental ratio.

I knocked on the door and he invited me in. By this time he was calling me Wil. He asked what could he do for me. My request for additional gas rations startled him. "We're down to only maybe another week of coupons," I said.

"We gave you a six-month ration. We don't usually give out supplemental rations. You are new and I'm going to make this exception," he said. "I'll give you the additional rations. Make sure to cut back on your driving."

After another two months, Emily again told me we were down to almost no coupons. I thought, *Oh no, not again.* She said she needed me to ask for about another month or so and then she'd cut back.

I went once again and knocked on the colonel's door. I told him I hated to come in again, but I'm here for another supplemental ration. He wanted to know what was going on. I had found out Emily shopped with the colonel's wife, the wife of the man I was talking to. I told him his wife, Joan, and Emily and three other wives were in our little Volkswagen bug. They wanted Emily to do all the driving, even though she was the lowest ranking of all the wives. The colonel understood, saying he got it, but this was our last supplemental ration so don't ask for more. I assured him he wouldn't see me any more about this.

I tried driving the English car I had been given within the housing community. On top of everything, it was a manual shift automobile, so I had to shift gears with my left hand rather than my right. I had to change my feet on the pedals, all of which took a lot of practice. I started driving because Emily and my family loved to travel.

In my assignment I worked for a great colonel with an exceptional team of officers. We worked on a sliding shift. On the early one, 7 a.m. to 3 p.m., I'd get off at 3. Maybe the next week I'd work from 3 p.m. to 11 p.m. The shift I dreaded was that last shift from 11 p.m. to 7 a.m. There were many times when I went home to rest, I found myself looking at the ceiling because I was accustomed to going to sleep at my usual time.

The wonderful aspect of this schedule was it allowed me to spend more time with my daughter. Emily had immediately found a job at one of the Air Force bases, Lakenheath Air Force Base. We had a nanny pick Lisa up from school when I was still at work and take care of her until I came home. Because of the sliding shift, I could pick her up myself many times.

We started reading to Lisa early. We would also have her read to us. She had about four or five girlfriends she'd gather and say, "Dad, here are three books. Will you read them to us?" The number increased to five or six picture books. They loved when I would read and point to the pictures. I spent a lot of time with her. Lisa had learned to read early. I was a member of the Literary Guild Society and I ordered books regularly. Lisa would intercept those books and read what she wanted to read, then slide them back in the box until I figured out what was going on. I'm almost certain her reading is the reason she was so successful in her later years of school.

Lisa was in a British school, Beck Row School, in Bury St. Edmunds. At her events, the teachers were constantly telling the kids to sit down, fold their arms and be still. Lisa was the only one who followed instructions. Many times I'd see her telling the other kids to behave. She was only four years old, but she was very mature because we trained her to pay attention and do as she was told. I give Emily all the credit for that. She really worked with our children. She wasn't a strict disciplinarian, but when she spoke, our children listened. I found out years later they said they could do anything and pull it off on dad because they said I didn't know what was going on. However, their mom knew what they were going to do before they did it and stopped them.

Wil Jr. adjusted quickly to life at Mildenhall AFB. We lived literally across the road from the base so had many amenities available for the those in uniform and their families. Wil Jr. joined other kids playing ping-pong, pool, bowling, or board games. He participated in the AYA athletic competitions, including baseball, football and track and field. He excelled, and I'd read his name, my name, in the base newspaper weekly.

I never had to spank my son; however he disagrees. If he touched something he wasn't supposed to like my stereo system, I would slap his hand away. They always said, "Mom would wear them out."

Wil continued playing basketball, baseball, and football. When he wasn't doing that, he would go to the bowling alley with his friends. We gave him time to socialize after he had studied. He did homework before he did anything else. He learned to be self-disciplined.

In his second or third year he was selected to be an exchange student in a British school.

In the UK tournament the Mildenhall All Stars were pitted again the other bases' teams including Chicksand, Bentwaters, Upper Heyford, and Lakenheath. All had their best 10- to 12-year-olds competing to get to the next step. As a parent, it's nerve-wracking to watch our kids. Wil Jr. was 11 years old on a team of mostly 12-year-olds, but the coaches had him batting first in the lineup and playing center field. Mildenhall advanced to the All Europe Tournament in Germany. When Mildenhall was knocked out, Wil Jr. told me, "We wanted to win so badly so we would get a chance to play in the tournament back in the USA so I could get a hamburger." These competitions were so enjoyable and rewarding for our family.

My boss in that first year, 1975, was Col. Robert (Bob) Pruitt, whose wife was Flip. My operations officer was U.S. Air Force Lt. Col. Carl Sanders, whose wife was Tad. I had a couple of very good friends, a Navy Lt. Cdr. Bryon and his wife June as well as Maj. Dave Harkavey, U.S. Air Force, and his wife Starr.

Byron Keesler was a White officer, lieutenant commander in the Navy. We were like brothers and to this day our friendship is strong. I didn't realize that he was three or four years older although we had the same rank. I was a junior major or an O-4. He was a senior O-4.

We were about 70 miles northeast of London and many times we'd go into London to do shopping. Emily loved to go to the shops. The British way of life was foreign to us, but they were very nice people. I enjoyed talking to the locals in the little pubs; they didn't see many people like me there. They ask where was I stationed. I'd tell them Mildenhall and they said, "Oh, you fly." The Air Force Bases around us were RAF Lakenheath, Upper Heyford, Chicksands, Feltwell, Bentwaters, and Alconbury.

In a typical week of flying, I would go on alert two days and on the third day, we would fly on a mission. When we came back, I had three days off. I spent all my time with my children and Emily. I could take Lisa to school. I could do whatever Wil wanted to do. Mildenhall was a wonderful family environment allowing me to spend time with my loved ones.

We had great opportunities to travel. The Air Force had an aircraft called the European Eagle, a big U.S. Air Force C-141 which departed Ramstein Air Force Base in Germany and flew into our base. If I was in a leave status I could take myself, my wife and my family free of charge to places like Athens, and spend several days there, depending how much leave I had. We could have even visited Turkey.

We decided to take a vacation to Athens. The kids were so excited to get on such a big Air Force plane. We did all the tourist-type things, the ruins, the Parthenon, and catacombs. On a boat tour, we saw the islands of Aegenia, Hydra and Poros and a lot more. Seven days and it was all about history. Many of the things the children were taught in school were in front of them in Athens. Our travel really supported their educational instruction. Our first year was full of enrichment activities and travel.

Young Wil was so good in baseball, he made the All-Star team at Mildenhall. He was one of the two or three young men selected to travel to Germany to play in the All-Star game. They didn't win the title, but what an experience for him to participate. I told him that when he was a baby, I was stationed in Germany and I spoke a little German. Here he was 10 years old and getting ready to turn 11. Emily and I didn't have an opportunity to go with him, but we had very good coaches who escorted the boys for a wonderful trip to Aschaffenburg.

In year two I was moved to a team commanded by Marine Corps Col. Duane Newton. It was the common practice of moving officers around. We worked very well together. We stayed in touch with him and his wife Joanne for years after leaving England.

During our second year, we did more traveling. My team had the opportunity to fly to Torrejón Air Force Base, Spain. During our two days, we took the crew van to sample the paella and sangria in Madrid. We even did a little shopping on the side.

We went to Rota, Spain and visited a submarine. I knew I was in the right service—the Army—as there was no way I wanted to be assigned to something so tight as a submarine. The sailors had "hot bunks." Sailors shared the bunk, so when a sailor was working, doing his job, the other sailor would be in that bunk sleeping. Touring the sub gave me a deeper appreciation for my Navy brothers.

Also in year two, we did some traveling with our best friends, Lt. Cmdr. Byron Keesler and his wonderful wife June and their two children, Lisa and Colin. Their Lisa was the oldest and she would sometimes babysit Wil and our Lisa when we all went off to the parties. We served on the same battle staff team and we became fast friends. We decided to travel to Italy in our separate cars because I wanted to take Emily and the children. We planned to visit Air

Force friends from our time in Washington who had been reassigned to Naples. We decided we'd follow the Keeslers in their Volvo in our Volkswagen bug, which had a four cylinder engine. Our plan was to drive through the mountains in Switzerland.

At Dover, we put our cars on the Hovercraft. Emily didn't like the ride, skimming across the water, but the kids loved it. It crossed the Channel to Calais, France, in 30 minutes. The cars were on the top and we were down in the passenger area.

We wanted to spend the night in a Swiss chalet. All the kids were having a field day. I was, too, because I had never done anything like this. I, a country boy from Goulds, Florida, in Switzerland, was thrilled. I had read about the Alps, but I never dreamed I would travel there. Those mountains were higher than high; I had to experience them first-hand to believe it.

After our night in the chalet, we were eager to get an early start. We learned quickly the proprietor had locked the chalet and we couldn't get out the regular door. So our son Wil and Colin climbed out the window and went to the proprietor's door to wake him up to tell him that their parents wanted to leave. They asked him, "Please be kind enough to unlock the doors so we can get out!"

Driving through the mountain peaks pushed the VW bug. I tried to pay attention to the road, not just the beautiful mountains. When I looked up, the Keeslers were gone! I finally caught up with them when they had pulled off the road at a scenic overlook. I told Bryon their Volvo was going too fast. My 4-cylinder bug couldn't keep up. He thought he was going slow. I asked him to please keep me in his rear view mirror.

As we were approaching Italy, young Wil said to me, "Dad, if I were you, I wouldn't change my dollars to Italian lira at the American Express office." I asked him why. I knew their offices were set up to accommodate American serviceman. He answered, "You can get a better exchange rate at one of these little Italian shops. They'll give you more lira for your dollar than American Express."

When I asked him where he learned that, he told me he had read it in a guide book. Sure enough, we got about 100 more Lira per dollar.

In Naples, we split up with the Keeslers. We found our friends, Burrell and Madelyn Handy. Wil was friends with their son Darrell from when we lived in Forestville, back in Maryland. We stayed with them for three days.

While there, I contacted the USO, an organization that supports servicemen and women when they are overseas; I asked them to plan a trip for us to tour key cities from Naples up to Venice. Our first stop was the ancient city of Rome. We spent three days and two nights and did all the sightseeing we could. The kids loved the Colosseum and the ancient ruins.

From Rome we went to Pisa to see the Leaning Tower. The kids had read about it in school. I couldn't believe I was standing in front of this famous tower. The kids wanted to climb it. I told Wil he was in charge, while Emily and I watched from below.

The USO plan directed us to go from Pisa to Florence. Florence was a very old city, and the food was very good. We laughed, "You can't get a bad meal in Florence." I knew if I ever came back to Italy, I wanted to visit to Florence again.

Our last stop was Venice with its streets made of water. We had to park the Volkswagen bug with its rack on top in the hotel parking lot across the street. We had made reservations at a fabulous five-star hotel, Hotel Bauer-Grunewald, which I believe stands today. All the other hotels, Roma, Pisa and Florence, had been two or three stars. I wanted to splurge because I really wanted to enjoy the city.

Apparently, the hotel personnel thought we had a lot of money because this was one of the most exclusive hotels in Italy. $85 a night! I whipped my American Express card out—I had been a member of American Express since 1968 after returning from Vietnam—to pay our hotel bill. Some of the best food in Italy was served in that hotel's restaurants.

Our two adjoining suites had a foyer in the middle. An attendant met us when we came home with a towel over her arm, and a jug and a basin of water to wash our hands and faces. She turned the comforters down and put chocolates on the pillows. The kids had the adjoining suite. We hadn't been separated before, so again I put Wil in charge. I told him, "Don't act up. Stay in your suite, because your mom and I are right next door."

Emily was always the explorer type. We traveled by gondola. We warned the kids not to fall out, but they couldn't care, they were having a ball. We had to take a boat ride to tour the Murano glass factory on its island.

Our last day, we left at zero dark thirty. We rode the gondola over to our car and put everything on the bug's top rack. When I look at a VW bug today—they are so small—I don't know how we did it. The kids didn't sleep one iota in the back seat. They were too busy looking at the scenery.

We took the northern route this time via Austria. We spent a few hours there so Emily could browse the shops and pick up a few souvenirs. My son loved to buy postcards and mail them to his friends whether they were back in the U.S. or in England to show them where he'd been. When we went to a hotel, he always looked for the room's postcards. Lisa would be right there with him, looking over his shoulder. She learned a lot from her big brother. We left Austria and headed back to the Calais Hovercraft to return to England.

In Dover, England, there was a long line of cars at the customs checkpoint. The red line was for people who had items to declare. If you didn't have anything to declare, you got in the green line which was just the opposite—no line. The rack on the top of the car was loaded with stuff we had bought. We pulled into the green line.

A British policeman came up and said, "Good afternoon, mate. Do you have anything to declare?"

"Sir, with all due respect," I said, "that's why I am in the green line."

"Are you certain you don't have anything to declare?" he asked.

I replied, "Sir, just a minute, I'm a U.S. Army major. We're here to help defend your country. If I tell you we don't have anything to declare, then we do not have anything to declare." I looked him in the eyes. He only said, "Okay, Major, come on through." Emily and the kids were cracking up in the back.

Wil said, "Dad, you can really get military."

After we got back on post, Emily and I decided we were going to take another trip to Lisbon, Portugal, but without the kids who stayed with our nanny. We spent three days and two nights in Lisbon and had a great time, enjoying a lot of old buildings and historical churches. Of course, we had purchased a guidebook and did our studies before we got there. We also went to Madrid. We bought a nice handmade ship and many Lladro porcelain figurines. Emily was doing all this shopping; there was so much she purchased in year two. I was just a young major with a certain weight limit for household goods that I could not exceed.

Prior to leaving England, my mother visited us. We sent her a ticket and flew her from New Jersey via Philadelphia, her first time flying. We picked her up at London's Heathrow airport and took her back to Mildenhall. She spent two weeks with us. My mother truly enjoyed time with my kids as well as touring all around England. This trip really touched me because it was a way to formally thank my mother for all the things she had done for me. We also extended an invitation to Emily's mom, but she didn't have time, or maybe she didn't want to fly the nine hours over the Atlantic Ocean. When my mom went back, she told all her friends she'd seen Queen Elizabeth. We did drive by Buckingham Palace and stopped the van, so she could get out and look through the fence. She didn't see the Queen, but it was a lot of fun for her.

In year three, I changed teams again because, according to my unit's leadership, I had done a great job on the other teams. They assigned me to the ground staff, which meant I only had to fly once a month, rather than four times, although I still received flight pay. I worked with Air Force Maj. Steve Gryzbowski, who became a very good friend to this day. He has unfortunately lost his wife Pat to cancer.

I read up on all the requirements of my new assignment. We administered all the standardized evaluation exams to the other four battle staff team members. I knew it was a prestigious position to be one of the two officers administering the exams.

My boss on that team was a Capt. Art Potter, a graduate of the United States Naval Academy. His wife was a beautiful friendly woman, Betty Potter. They lived behind us on Colonels' Row. I believe the reason he selected me was my background in math and how I done so well on all team exams during my first two years.

Capt. Potter has gone home to God. I am now the same rank as he was, but if I met him today, I would still call him captain and Sir because of the respect I had for this man. We made wonderful friends on that team. Maj. Charles Kitchell and his wonderful wife Charlotte and Army Maj. Dave Fitz-Enz were also on the team and I consider them all dear friends.

We also had another set of visitors, my sister Brenda and her daughter Thalaya, and my first cousin Clara White and her daughter Selena. They spent a week with us. They had fun doing all the tourist things, seeing the ancient historical sites. Our third and final significant visitor, a friend from Emily's past, was Army Lt. Cecelia Hunter. She was a commissioned second lieutenant in the Army and was stationed in Germany. She flew over and spent several days with us. Emily babysat her when she was a little girl. Cecelia also babysat Wil Jr. when he was a baby. Her young sister and brother participated in our wedding in 1963 as flower girl and ring bearer. To this day, this family remains close to us, especially Cecelia. When she was growing up as a young girl, she and her brother Byron and I would play tennis when I visited Savannah. Bryon would retrieve the tennis balls for us when they went astray; quite a few did.

In my last year, I placed a down payment on a small convertible sports car, an MG. Another friend of mine on the team had one; he convinced me I should get one. I went down to the salesman right on the base who was selling the MGs. I told Emily about this transaction before I took off on my weekly flight. When I got back from my mission, I walked by the place where I put the down payment.

The salesman called me over. He was hot; very mad. "Your wife found out about your down payment on the MG, and she decided to take it back. Are you the man of the house?" I told him I was, but I listened to my wife.

When I got home, I asked Emily what was going on, she said, "Only two people can ride in that car and there are four of us. We can't buy a car that only two can ride in. It must be big enough for all four of us. And that's that!" I only said, "Okay."

That salesman remained mad at me, saying, "Here in England, men make the decisions." I had to tell him in America we didn't operate like that.

To this point in my Army career, this was my best assignment: flying all over Europe, giving my family an opportunity to travel and experience sights we'd never even dreamed of. It was especially great for my two children because, like my mother, we put a lot of emphasis on education. Many subjects taught in school, Wil and Lisa had a chance to see firsthand, especially in Germany, Italy, and Spain.

Near the end of our assignment, I knew I was going back to Washington for my first Pentagon assignment in the Defense Intelligence Agency, located in Arlington Hall Station. Also I had to get us a place to live on our return to the United States. In April 1978, I was able to hitch a ride on a KC-135 aircraft that was being returned to the U.S. for overhaul. I was one of three passengers on this KC-135 Tanker, flying into Pease National Guard Air Force Base in New Hampshire. It was an eight-to-nine-hour-flight. We landed there and I was put in a limo and taken to Logan International Airport in Boston to catch a flight to Washington, D.C.

In Washington, I was picked up by Erma Jones, the wife of one of my best friends. Col. Lavert Jones had been my classmate. Emily and I already did our research based on the real estate information sent by the Jones.

This house would be the first home we had ever purchased as a family. Homes were very expensive in Northern Virginia. On the first day, Erma showed me all around. We found the place we wanted to

buy, in Burke's Centre, Virginia, which was a new community being built with several subdivisions. In one subdivision, Spring Pond, I fell in love with one of the colonial homes, but when I looked at the price tag, I thought, *Wow!*

I called Emily to tell her about it. I told her I had done the math. Even though I knew she would get a job. I wanted to get this home based on my salary. We had good savings. She told me to make the decision based on what she and I had decided we needed—four bedrooms, at least two and a half baths and a two-car garage. We had to have a place to stay, and she trusted me to make the decision. The house I wanted was $91,000. We had to put down $20,000 as the down payment. I went the next day to see the salesman and completed the paperwork. I called Emily and told her we had a home.

In August 1978, we rotated back to the United States. We had to say good bye to all our friends, but that's the military—you make friends and then you leave them. It was especially difficult to leave the Keeslers. We shipped our car, the VW bug, from London where it was placed on a ship to be sent back home. We also had ordered a Mercedes 280SE.

I was so proud of our first home. Each of the children had their own bedroom. We had a nice master bedroom with a fireplace. It was a great community located at the end of a cul-de-sac with wonderful neighbors. Our house backed up to a pond. Lisa was an explorer, too. At that time, she was six years old. She was so bright and was scheduled to go into first grade, but when they tested her, her placement scores were so high, the counselor wanted to skip her to third grade. We said no to that big a jump. We compromised and put her in second grade. She was reading at a fifth or sixth grade level at that time. She was also strong in math and social studies. The British school system had really prepared her, giving her a solid foundation to take off academically.

Chapter Eleven

The Pentagon and Helpful Neighbors

England had been my best Army assignment not only for my family but for my career. I received three years of outstanding Officer Efficiency Reports. This OER is the lifeline for any officer. If you don't get an outstanding OER, chances are you don't stay in the Army very long. At that time I was in a Voluntary Indefinite category, meaning I continued to re-enlist and serve, but I was still a U.S. Army reserve officer which was my designation coming out of college. In my view, it's very difficult to get in the regular Army officer status. I had applied in Germany to get a regular Army commission after I received my first two outstanding OERs but was turned down; Headquarters Department of the Army never gave me the reason for their disapproval. I thought, *That's fine, I will continue to perform the best I can.*

In Washington D.C., many of my friends tried to figure out how I maneuvered my assignment to Defense Intelligence Agency. As an U.S. Army reserve officer, I could only stay on active duty for 20 years. My alternate specialty training allowed me to get this key job in DIA. I had told my friend in officer assignments I wanted to get an assignment in Washington, D.C. He delivered and that's how I got to Washington.

My plan was to come to the Washington D.C. area because it would afford me more opportunities to get a high-salaried job once I left the military. That was my plan of attack; only Emily and I knew about it. I would reach my 20 years of service in 1982. We rotated out of England in August 1978, which gave me four years to make contacts for the transition to civilian life.

I was happy to be returning to the United States; although we had seen my mother and my sister, I wanted to see Emily's mom and the rest of our families. Family motivated me to be the very best officer I could be.

At this point in our marriage, I had learned a lot from Emily about networking, connecting with people. In the military as an Army officer, social etiquette went hand in hand with your assignment. There were many social functions to attend. If you slipped up and didn't do or say the right thing, you could find yourself with a bad OER, even though it would be veiled under something else.

As a reserve officer going out at the 20-year mark, I didn't want to do anything wrong. I bought and studied several books on Army officers and officer protocol and etiquette. As officers, we had calling cards with our name and rank. When you visited a senior officer's home for dinner or for social cocktails, it was proper to leave your calling card at the officer's home. I cautioned Emily because she didn't know very much about the military. Further, she didn't care about the rank because she saw people as individuals, man or woman, which was good in a lot of ways. But I'd always remind her that's the colonel or the general and I'm just a young major.

In our new home in Burke Centre, I never dreamed I would purchase a home of that value. I think back to the home my mother built. She didn't pay more than $12,000 to build our home.

In August 1978, in our cul-de-sac, all the families had young children. The family next door, Al and Carol Stolpe, had three little girls and Lisa loved them. Wil was too old to play with those girls, but there were other kids in the community to become his friends. Down the street was a basketball court, which was one of the reasons I chose that community with all its amenities for families.

I left early for work, but I never worried because we had wonderful neighbors. If Lisa got off the school bus before Emily came home, she'd have called one of our neighbors to watch for her. Lisa also had a key around her neck and knew to go in the house. Our dentist even lived in that community. He ended up being our dentist until he retired from his practice.

When we moved in, we were the only Blacks on that street. One day, one of my neighbors told me another Black officer lived on the street one over from us, Col. Colin Powell. I had heard about him—he was a rising star in the Army. Little did I know he would go on to be chairman of the Joint Chiefs of Staff, then U.S. Secretary of State. As just a major, I wouldn't dare go over to introduce myself.

One day while I was mowing my lawn, a gentleman pulled up in an old Volvo and got out of his car. I found out later he loved Volvos, working on them and restoring them. I stopped the lawnmower. He was a tall man with a broad open smile.

"Hi, I'm Colin Powell. My wife is Alma and we have three children. We have two daughters and a son." Our children became friends.

"It's my pleasure meeting you, sir."

"I'm going to be promoted to brigadier general soon. We'd like to invite you and your wife to come to a small reception at my home. I would like to introduce your wife to my wife."

"I look forward to it. Once you get the notification, we will be delighted to attend."

Over the next months, I had the chance to get to know him. I knew he was serious as a judge at work, but I found him playful with his family and children, and good to his neighbors. During those days and even to this day, senior officers don't socialize with junior officers that far down the "pie chart."

Brig. Gen. Powell and his wife had a small reception for about 30 people at their home. I felt honored for Emily and me to be included. During that time in the military, it was not common to see someone of our rank at a general officer's home for an event such as this. As I recall, the great majority of the assembled guests were courteous, but curious about who we were and how we were in invited to such a gathering. Of course, we mixed well with the military and civilian guests at the reception. Since it wasn't on a military installation, I felt at ease after the first few minutes there. Emily was always charming and could mix well with anyone, regardless of their station in life.

At the DIA, I was assigned to a team with retired Air Force Lt. Col. Sy Simonson. He became my best friend in that office. Every day around lunchtime, he'd say, "Wil, let's go get a cup of soup." We'd take a short walk to the officers club. I didn't like the assignment because of the sensitivity of the job in intelligence at the highest levels of the Defense Department. We only wore uniforms one day a week, which was kind of strange to me, but the reason was to make sure our uniforms still fit. We had to remain fit because I was in the Light Infantry and I had to be ready to go to war in the event something happened overseas. Working there, Sy Simonson taught me the procedures and how to navigate the DIA environment. Looking back, my time at DIA was a unique opportunity.

If I drove my car every day, I would have a 45-minute commute. I joined the carpool with Navy Lt. Cdr. Charlie Bradner and a Department of the Army civilian. With three of us, we could ride the HOV lanes on I-395 going into Washington. Also I had forgotten about the cold, snowy weather in the D.C. area.

The blizzard of 1979 was a bad snowstorm. The federal government was always late in making the decision to close. Of all days, it was my day to drive the car pool. I had rear-wheel drive in my Mercedes-Benz. A Mercedes-Benz is a nice car to drive, but rear-wheel drive is no good in snow and certainly nothing is good on ice.

Around three in the afternoon, the snow was really coming down. When we received word that the federal government was closed, we three hustled to my car. We were able to get on I-395. Cars had stalled and were skidding into snowbanks all along the highway. Numerous accidents had occurred along the route.

I don't know how we made it up to the mixing bowl interchange on I-395, but we did. We got off on Old Keene Mill Road, which would take us directly into Burke. I told the guys, I didn't know how far we could go in this car. We might have to abandon it and walk the remainder of the way home. We got close to the intersection of Rolling Road, but it was hilly. All we did was spin. We couldn't get traction. After spinning into snowbanks a few times, we pulled the car over in some soft snow. It was a dry snow of about eight inches, not a wet snow, which was good. We were able to pull into a shopping center and park my car. In the Army Infantry I was always prepared for the worst. In my car's trunk, I had a pair of boots. I was in my Army Green uniform with dress shoes, so I changed them for my boots, put on arctic mittens, a cap with a flap, and an overcoat.

We joined about 300 people walking along Old Keene Mill Road. It took us an hour to get to Burke Center. The Navy commander was from the Northeast, so he was acclimated to snowy icy weather. The civilian was also from the Northeast. I, the lone Florida boy, was the only one suffering. What an experience! Two days later after they lifted the suspension of work, I was able to go get my car from where it was parked.

In my family circle Lisa was growing so fast. In the early '80s we met Col. Tony Manning, USMC, and Marcia Manning and their three children including a daughter, Nicki. We, including Lisa and Nicki, became good friends. At the Mannings' invitation, Lisa joined their daughter in the Jack and Jill Club, which worked with young kids to put them in leadership positions, emphasizing education and etiquette. While the country was integrated, folks were still separated in social activities, so Jack and Jill provided great fun for the kids.

In 1978 young Wil was active in school with football. Wil had played three sports in England. At West Springfield High, a school of 2,500 students and only 40 Black students, Wil Jr. ran into some challenges. Because he had been advanced a grade, he was one of the youngest in his class. He was only 5 feet 1 inch tall and weighing only 120 pounds . When we got to West Springfield High School, the football coach for some reason decided he wasn't going to play my son. I went to a few games. I knew my son could play. I told Young Wil I was going to speak to his coach. He didn't want me to do that. Emily also thought that was a bad idea because I might say the wrong thing.

Against his and Emily's advice, I made an appointment to speak with the coach about my son's lack of participation on the team. I left the Pentagon to meet the coach. When I got there, he was nice and courteous. Keep in mind, this was a predominantly White high school in Virginia. Wil Jr. went out for the freshman football team against boys 14 and 15 years old. He made the team but was no longer the star. The coach made him back-up quarterback and put him on special teams for punting.

An older student on the varsity squad, Jack Brooks, was already 5 feet 10 inches and 180 pounds. Wil Jr. asked me to buy him ankle weights and a weight set to use to get stronger and jump higher. Wil Jr. was never discouraged; he viewed it as incentive to work on his game.

I told the coach I wanted him to give my son a chance to play to show him what he could do. I mentioned he was a running back, but he could play several positions. I knew the coach had a running back on the team, but my son was second-best to him. All I asked was to give him an opportunity to show his skill set. I reminded him that I pay taxes in this county. I explained my high school athletic career and how important to a kid these experiences are. He agreed to give Wil a chance. The next game he put my son in and the rest was history. He didn't start, but he played. That's all I wanted. Wil was a great football player.

Homecoming at Wil's high school was rapidly approaching in the fall of 1978. I asked my son if he was going to the homecoming dance. He said he didn't have a date. He didn't think he wanted to go. I told him as a football player, he was part of that school. I didn't want him to miss this big event in his high school life.

In England, Wil Jr. had always been a socially involved kid. Many times I had taken him and his buddies to the many dances at school and it was my time to bond with him and his pals. I would ask them who had the best dance moves and who had the best "program to get the dollies to dance with them." My son and his friends would be cracking up. Even in their 60s they are still friends. I argued Wil Jr. should participate in the social life of high school, too. But there were only 10 Black kids in his grade and he didn't really know any of the girls.

I talked to my friend and he agreed Wil could take his Heidi to the homecoming dance. They were great young kids. Wil's friend Riley Hamilton took another young lady. They couldn't drive yet, but I offered to drive them to all the events. Before homecoming, traditionally the boys would take their dates out to dinner. That was no problem. I drove them to Washington, D.C. down to the Waterfront. I gave my son my credit card and dropped them off at restaurant called Hogate's, a seafood restaurant in southwest D.C. I drove them back to the dance. I told them I would be here in the school parking lot to pick them up at 11 p.m. when the dance was over.

The period I was assigned in DIA was active socially for us as well as for the children. We started a Christmas social with about three other military families. We gave the first one in our home. We would prepare the food and we exchanged small gifts on Christmas Eve. Every year we would rotate to another family. The next time it was at Col. and Mrs. Robert Warren's home and next Maj. and Mrs. Riley Hamilton's.

I had only a brief period in DIA. One day, I was called down to the colonel's office to hear some good news. I had been selected for promotion to lieutenant colonel. I was so surprised and pleased I wanted to jump out of my suit. He further said I had been selected for 0-5 battalion command. I would command a light infantry battalion in the United States. I couldn't wait to get home and let Emily know the good news.

There was a high probability that we would be moved soon. Things started to happen fast. I received a letter from the Military Personnel Center, Headquarters Department of the Army. The letter congratulated me for being selected for battalion command. It reminded me also that I was a U.S. Army reserve officer, which means that the 20-year mark of my career, I would have to retire unless I applied for a regular army commission. If I received it, I could stay at least another 10 years.

So, in a subtle way, the United States Army asked me to apply for a regular Army commission. They had earmarked me as one of those officers going up the ladder. I applied for the commission and in record time—about three weeks—I was approved for a regular army commission. They immediately swore me as a regular army officer. I was on my way. I owed it to the three years I spent in England, with all outstanding officer efficiency reports and my year In Kansas. Also there were four years at Howard University, all with outstanding officers efficiency reports. I had eight successive years of outstanding OERs, which put me in a different category. It appeared I was one of those officers earmarked and designated for a higher level of responsibility and promotion.

Emily was asking questions and the kids were excited that Dad had been selected for another assignment and for promotion. Young Wil wasn't in attendance, but Emily and Lisa were there at my promotion ceremony to lieutenant colonel. Emily participated and took my major's leaf off and pinned a silver leaf on one side while the general pinned the other leaf on the other side. I was a regular Army lieutenant colonel, which meant I was ready for bigger and better things.

A letter came from the HQ Department of the Army, which told me I would be sent to Fort Knox, Kentucky. I was going to command a mechanized infantry battalion. I really didn't want to go to Kentucky, but command is command. My oldest brother Alvin and all my Army friends had told me 0-5 command is very difficult to get—you are in a small group of Army officers who get the opportunity to command. Emily was also not very excited about going to Fort Knox, Kentucky. She wanted to know what was there. I told her

101st Airborne Division was located there and Fort Knox has all the gold. She was still not excited.

Two or three weeks later I received a call on my direct line from Headquarters Department of the Army Assignment Branch. They decided to change my orders, sending me and my family to Fort Ord, California, which is close to Monterey, Seaside, and Carmel. I went home and told Emily, Wil Jr. and Lisa we were going to California. We were all delighted.

I was slated to assume to command of the second battalion of the 32nd infantry, which was a light infantry battalion of 737 men. I would be assigned to the Seventh Infantry Division there at Fort Ord.

During those days, many Army personnel wanted to be assigned to either Fort Carson, Colorado, Fort Ord, California, or Fort Louis, Washington. The weather, the fresh air, and the Pacific ocean! This assignment was a big plus for our family.

Chapter Twelve

California and Command

On the way to California, we drove a southern route through Louisiana and Texas to see friends from England who had relocated to Dallas-Fort Worth. Emily also wanted to visit a friend in New Orleans from her Tuskegee dietetic internship program.

We had a golden plan to entertain the children on the drive with a small 5-inch TV. We plugged it into the cigarette lighter and we set it on the center armrest. We hadn't thought about reception. Once we left the Washington, D.C. area, all we had was snow on the screen.

Our first stop was New Orleans. I'd traveled all over the world, but never this fun city. Next, we spent the night in Plano, Texas with Charlotte Kitchell and her family. She and Dale Kitchell, an Air Force officer, had separated by that time. Her husband taught me so much about communications. I called him and we caught up. Our drive took us a leisurely four days. Along the way, Wil sent hotel postcards to his friends.

Emily's Volkswagen bug was driven by a driving company across the U.S. It was one of the best cars we have ever owned. With over 200,000 miles on the odometer, it had literally traveled all over the world and was practically a member of the family.

When we arrived at Fort Ord, it seemed like everybody was waiting for us. They put us into the bachelor officers quarters for two weeks until my commander's quarters could be ready, which turned out to be a nice, three-bedroom home at 314 Metz Road, right on the post.

At Fort Ord, I reported to the division headquarters. I had to pinch myself. I was so excited to command, which was the most important job thus far in my military career.

The 2nd/32nd Battalion, as a light infantry division, must be prepared to go to war on a moment's notice. I had three company commanders and a combat support company and also a headquarters company. It was in the 7th division, commanded by Maj. Gen. Thomas D. Ayers, a balding and bespectacled officer and a nice individual. Everything we did in this unit was under close observation. I asked my family to understand that command was serious business because it was going to take me away from our usual family routine. I felt blessed that Emily was my partner. She always looked to the lighter side to keep everything even keel. She provided me support as she understood the job's importance.

My change of command was an impressive ceremony. The division band performed. The outgoing commander and his staff lined up. The unit's colors were given from the outgoing commander to me. I took the battalion colors and gave them to my flag bearer. The two of us then trooped the line. We passed in review of the battalion's soldiers saluting the outgoing commander, saying goodbye to him and welcoming me, the new commander.

After I was sworn in, I had a meeting with my staff to introduce myself. I wanted them to hear directly from me and shake my hand. At that time, I let them know what I expected of them and established my standards. If they didn't want to work hard, I advised them put in a request for a transfer and I would accommodate them. I wanted to have the best battalion in the 7th division.

My brigade commander, Col. Don Chunn, was a wonderful, blond-haired, blue-eyed guy who was personable, low-key, and smart. He was a man of the highest character and intelligence. I knew right away he would provide me the support I needed to do my job. Luckily I didn't have to make many staff changes in the battalion. I had outstanding officers with a mixture of West Pointers and ROTC officers from colleges and universities around the country. A military academy graduate commanded one of the battalions in the brigade and the third commander was a product of the California higher education system. We were all friends, but we were very competitive.

My son attended Seaside High School. He lettered in football, baseball, and basketball. He held class offices. Even as the new kid on the block he found his place all the while he was preparing for college. The primary thing—he was an academically strong student. He smiled more easily than me, totally different from his dad. My brothers had warned me you're going to be in the infantry, just don't go around smiling but have a poker face. It would keep the others at bay and not allow them to figure you out. That was their approach, but I found that wasn't me.

Lisa was at Fitch Elementary School, which was a big plus. I found out that one of my friends from my Fort Jackson, South Carolina days was the assistant principal at that school. He'd retired from the Army as a major and gone back to school to get his master's and then his E. ED in educational leadership. I didn't tell Lisa right away because I did not want my friendship with him to be a crutch. Later, I let her know that Dr. Laster was my friend. He was senior to me by two or three years. I never undermined him as the school administrator.

Emily, who has always worked, got a job in the Salinas School District. However, she maintained her responsibilities as the battalion's First Lady. All officers' wives in the battalion came under Emily's area of responsibility to organize spouse activities on a monthly basis. Little did I know, her leadership enhanced my performance because wives or spouses talked to their spouses about Mrs. Bryant.

A father lives through his children, especially sons when they play sports. I always told young Wil, if he caught a pass or intercepted the football, I wanted him to take it to the house, which means score a touchdown.

Seaside High School primarily accommodated military kids. Fort Ord was a large infantry division with over 15,000 soldiers. Almost all these men and women had children, so probably 80 to 90 percent of the students at the school were military brats. Military brats are tough kids who have traveled around the world and can easily acclimate to a new environment.

I was a hands-on kind of father, but Wil reminded me I was not a battalion commander of their games. I needed to stay in the stands and let him take care of things down on the football field. In one game, Wil made a hard tackle, causing him to appear unconscious on the field. I ran out of the stands down to the field. By the time I got there, the trainers had smelling salts under his nose.

When he opened his eyes, he said, "Dad, what are you doing on the field? Get off the field." Then I knew he was okay. I ran back to the stands. Wil later said, "Don't ever do that again. I'm not a small kid." I told him you are still my son, no matter where you are. Wil Jr. led the league in interceptions and defensive touchdowns. He did

receive offers to continue playing at the college level but he had zero aspirations to do so. Wil Jr. had read that the average NFL players only lived to be in their 50s and even at 16 he was thinking long term to improve his chances of a full life.

At the homecoming game against Monterey High School, fans packed the bleachers and the end zones. In the fourth quarter in a tight game, the quarterback reared back and threw the ball. My favorite player for the Washington Redskins was Charlie Taylor. Every time he scored a touchdown, he held the ball up and gave it to the official. If he ever caught a touchdown pass, I had told Wil to hold the ball up like Charlie Taylor for me. Wil intercepted this ball. Everybody jumped up in the stands. He ran down the sidelines—he was very fast. When he got in the end zone, he turned, and he spiked the ball. Touchdown! The referee gave his team a 15-yard penalty on the kickoff for his unsportsmanlike conduct. I almost jumped out of the stands—I was the proudest dad in America. Seaside High school beat Monterey, the bigger school, in the homecoming game.

Lisa had learned to play the piano my mother had purchased for us. Part of my job when I got home was to take Lisa to her piano lessons taught by one of the officer's wives. It was very expensive. I was paying $35 for half an hour in 1979. At her recitals, she insisted we not make any noise but wait until we were told to applaud. She played to perfection. It warmed me and my mother's heart when I conveyed to her how well Lisa was playing.

At the battalion, my group of young officers did a great job. My executive officer, Maj. Brook Harris, had an iron hand with the officers. He knew how to get the very best out of them. If an officer wasn't performing, he'd tell me who was not suited for a position, and he'd recommend moving him into a line unit. I'd tell him, "Execute. Carry it out." That's how we did it.

My command sergeant, Jerry McCullough, was a master parachutist with over 70 jumps. He placed the right NCOs in positions, ensuring they were highly competent in their jobs. The NCOs had to know the ins and outs of all their equipment, weapons, generators and vehicles.

Another NCO, my mess sergeant, was in charge of our consolidated dining facility, where over 700 people ate their meals. He knew his job, but I told him about my experience with cleanliness and nutrition in food service in Germany. I mentioned my wife was a nutritionist, although she would have no role there. He needed to be ready when the division inspected us. Our dining facility was judged the very best in the 7th Infantry Division, receiving the Leo Cobb Memorial Award, the top award for a consolidated mess.

My first Thanksgiving in the battalion was a big day as it is in all the military units, so different and upbeat compared to my Vietnam Thanksgiving. All the families are invited to come to the battalion or whatever unit to eat with their spouses. I wore my dress blues. All my cooks in their clean white attire stood at ease with their hands behind their backs. My battalion mess officer and the mess sergeant greeted me at the door, saying "Sir, the battalion consolidated facility is now open. We are ready to serve you and the troops."

I told them to let the NCOs and the command sergeant majors go ahead. The mess officer interrupted me: that would break protocol because I, my wife, and children should be the first to go through the line. I gave my kids "the eyes." They had me really on edge. I wanted them to be perfect children, but they always asked for more. Lisa, cute little Lisa, always wanted seconds. I told her, we have a saying in the battalion, *take all you want, but eat all you take.* She pleaded with me about seconds, so she got her way and had a clean plate. It was a wonderful day, etched in my memory.

In California I was deployed for three big deployments of training. As a light infantry battalion we had to be prepared for jungle, cold weather, and desert conditions. I traveled with the entire battalion to Camp Roberts, California and to Fort Hunter Ligget and to Fort Irwin, California, the National Training Center on a regular basis.

At Fort Irwin, in the Mojave Desert, was the ultimate test, we trained to fight like it was the Middle East. We had to survive in the intense heat, which affected all our communications gear. If I needed to reach the west side of the mountain, I needed a relay site to communicate from the east side because those 8,000 to 10,000 foot mountains blocked the signal. We called for a helicopter to put some radios on their highest mountain, which was a barrier to my units on the other side. Up there a team of soldiers manned those radios.

Before we went to Fort Irwin, the command sergeant major assembled the entire battalion. First, I spoke to the battalion about our 18-day deployment. We needed to learn about conservation of water and how to keep our equipment running, so we could be ready to fight when called.

The sergeant major, the top NCO, next told them, "Men, you are going to love our training at Fort Irwin because there is a woman behind every tree there. You'll get a chance on your time off to have a lot of fun."

Our long convoy of Jeeps and trailers drove to Fort Irwin. When we arrived, we all saw there was not one tree in the Mojave Desert—it was all tumbleweeds.

Temperatures rose to 110 to 115 degrees during the day and at night dropped down to 40 degrees with winds.

My troops had two-man pup tents. I, the battalion commander and my driver, had a Hex tent with two cots and a small stove. The first night after my driver turned off the lantern, we heard *sssshh*.

Pigmy rattlesnakes! I hate snakes. One had squirmed into the tent. We tore the tent down and I never slept in a Hex tent again. Instead I slept in my trailer. I had to make a communication check with the brigade HQ every hour. I had my driver tie down the top after I got in with my radio, so no snakes could harass me. I swore my driver to secrecy about me being in the trailer. The only thing in the world I'm afraid of is a snake.

Two more great sets of training opportunities pushed us to the top in terms of combat readiness. The first was jungle operations, where I took the entire battalion to Panama for 16 days of training. We had the chance to tour the Panama Canal before the U.S. government turned it over to the Panamanians. I was driven into an empty lock in my Jeep, followed by some of my staff. It was quite a sight, the concrete walls and massive lock mechanism.

Somehow the commander sergeant major found out I was afraid of snakes. One night when I was getting ready to retire, he called and asked me to come to the command post. There three of my largest soldiers wrestled a 12-foot boa constrictor. It appeared to be a foot in circumference. My feet automatically took off. I told them to get rid of it, take it back to the jungle, I didn't ever want to see it again.

The training required us to do a three-day exercise living out in the jungle at the end of the field training. There was no way I was going to sleep in my Hex tent. I had my driver prep my trailer the same way I did in the Mojave Desert, so no snake would get in.

The other significant training was Cold Weather Training at Camp Ripley, Minnesota with the National Guard for 16 days. We flew into Minneapolis-St. Paul and many of the soldiers, including me, even though I had been in Korea, had never experienced cold weather as we experienced there. Nothing was as cold as Camp Ripley in February with a 38° below zero wind chill factor and a foot and a half of snow on the ground. A convoy of Greyhound buses took us to the training location. Our weapons required special treatment, the right kind of oil, or they wouldn't function. Our trucks and our vehicles wouldn't run unless we heated their engines. We wore layers of cold-weather gear. We had to learn how to fight in the snow. We were taught to ski on the bunny slopes and to wear snowshoes for cross-country movement.

After we'd been there about seven days, I needed a haircut, along with several of my officers. We drove into Minneapolis. We stopped at the first barbershop we found open on Saturday morning. I walked in first as the leader of the group. The barbers, who were all Caucasian, had probably never cut any Black man's hair. They looked at us—surprised. I told them who I was and all we needed was to get our hair trimmed off our ears. Everything was very tense until I told them I was sure they could do it. They said, "Okay, you're first." My guys got a big kick out of this barber kerfluffle.

At Camp Ripley I had a task force of approximately 1,100 soldiers; during our stay I did not have a single disciplinary problem. The only problem we had was trying to keep these troops warm. On the day we were scheduled to leave, when the convoy of buses arrived, their cheering could be heard from Camp Ripley to Los Angeles. They were so happy to get out of the cold. I felt the same way, but I couldn't say anything.

Approaching my officer's efficiency report, I thought I was in a good position. We had the best consolidated dining facility, and we had the best reenlistment numbers. We had the Army Training Evaluation Program, where the entire battalion was put out in the field and another unit evaluated us. We performed very well. The big exam was the Command Maintenance Management Inspection, of all of our equipment and wheeled vehicles. I had close to 100 vehicles in that battalion. We scored very high, so I knew that I was on my way to my first outstanding Command Level OER.

The OER determined selection for promotion to the next grade and for the service schools. During those days the Army's OER used three ratings: a top block, the second, average; and the third block, below average. If you end up there, you're done, especially in command. I knew I should be in the top block, based on the performance of my battalion.

When I received my first officer efficiency report, the division commander, Maj. Gen. Thomas Ayers had rated me just below the top block. I couldn't believe it. No other battalion in the second brigade had done what we had done. I went immediately to see the brigade commander and expressed my displeasure. He thought it was a very good OER, above average. I argued we had done top block work. Rumors circulated that the commanders of the other battalions in the brigade, had received a top block. I told my brigade commander I was going to request an appointment with the division commander and ask him to change my rating. We had beat the other two battalions in everything. My commander told me to go ahead. For the appointment, I sent up a one-page summary of all my accomplishments.

For the appointment I walked in, saluted, and identified myself. He listened to me and reviewed my one page. He told me, "You are an outstanding officer. This is just your first report. I'm sure if that you will continue to perform the way you're performing, you'll find yourself in that top block the next time."

"Sir, with all due respect to you, if I get a nick on my OER like this, it is going to impact me for selection for senior service colleges and on my promotion to full colonel, possibly my promotion to brigadier general. He agreed and also disagreed. He wasn't going to change my rating, but he believed I'd get the top block on my next OER if I continued to perform at this level. I thanked him for his time, but obviously, I was unhappy. I thought he was playing favorites because this other officer was a West Pointer, and I was a ROTC type. The West Point commander of that battalion, a good friend of mine, ended up a three-star general.

Shortly after that report, we received a change of command of the 7th Infantry Division. A new division commander, Maj. Gen. Thomas Moore, took over. He liked my battalion when he came out and surprised us for a quick inspection. We did a great job.

Again, I thought my battalion performed better than all the other battalions. And we were selected to travel to Germany to a major annual exercise called the Return of Forces (REFORGER) to Germany. We packed all our equipment up, trucks, weapons, generators, and shipped it to Holland. We flew to Germany for 30 days of training. My battalion, 2nd/32nd, did so well, we were selected to train with the German Army at Wiesbaden. I was ecstatic because I thought we were the very best.

The German training exercise also gave me an opportunity to go back to Würzburg and find my landlord, Herr Fleckenstein and his wife Frau Fleckenstein.

On my day off, I walked into his print shop and asked the proprietor, speaking in German, if Herr Fleckenstein were there. Out came my former landlord, unchanged except a bushy mustache. He stopped in his tracks and said, "1st Lieutenant Bryant!" He shook my hand and gave me a hug. By this time he spoke a little more English now. We had left Germany in 1967 and it was 1981. I asked about his wife, but they were divorced. His son was away at school. Of course, he and my son had played together. He asked about Emily. I got the news about our other friends, Hans and Barbara Klorters; they had also divorced. We had all socialized together; I'd practice my German with them. We both shed tears and he shook my hand good-bye.

The next week, my battalion and I returned to the States under the colors of a job well done. I received my next OER, and I received a top block. The division commander was pleased with our performance.

One day while I was sitting in my office, the command sergeant major called me to the phone; Lisa was calling. She begged me to come home right away. I asked her what was wrong. In those days, like in Burke Centre, Lisa would come home alone before Emily arrived. She had a house key and other families close by would look out for the kids.

"Dad, they are painting our house pink. I don't want a pink house," she cried. I asked her to settle down and I told her the post maintenance people know what they're doing. She insisted, saying "You said you're the commander. Come home and make them paint it something else."

I went right home. The paint was deep orange, not really pink. I told her not to worry about it. I'd take care of it. I also explained orange was okay. That's how we got out of that fuss.

One final deployment near the end of my 30-month battalion command was to Singapore. We had done so well that the division selected the 2nd/32nd infantry commander and his staff to travel to Singapore to train with the Singaporean Army for 16 days in October into November 1982.

Singapore was one of the most interesting cities I have ever visited. It was the cleanest and safest city I've ever been in. There was a story about an American who threw a cigarette on the ground and his punishment was 11 lashes in a public venue.

One night I went to dinner with one of my staff. At 1 a.m. we had to walk one quarter mile from the restaurant back to our lodging. I never had to worry about crime—not one iota.

At Fort Ord, I had some of the finest officers in the United States Army. We trained them to do their job efficiently. I treated them with respect and they performed for me. I gave them outstanding efficiency reports because they earned them.

By this time Wil was set to graduate from Seaside High School. He earned a four-year Army ROTC scholarship so he could attend any institution that had an Army ROTC program. All his tuition and fees would be paid for.

We asked him what were his top school choices. He wanted to attend the University of Southern California. Emily, more than me, wanted him to attend Notre Dame. I thought about the logistics—we would have to fly to South Bend, Indiana. The weather was very cold there, but I didn't want to overtly influence his decision. I wanted him to make his own decision. He chose the University of Southern California.

In June 1983 at Seaside High School Awards Day, Wil would be presented with the Army ROTC four-year scholarship. I was invited out to make the presentation. I felt like a big man coming to that high school, and I was as proud as any dad can be. In front of the entire student body, I made that scholarship presentation to my son. I also wanted to dispel any doubts that my rank as Army lieutenant colonel affected his selection; on the contrary I had no impact on that scholarship. The selection was based on academic performance, SAT scores, and passing a physical exam. His SAT scores were very high. Academically he had performed very well.

I took him down to USC for freshman orientation. By that time I had received my orders. I was to be reassigned to the Pentagon. Emily had asked me to take him to orientation, and said she'd see him when we headed back east to our next assignment. Wil was going to be living in Touton Hall, which was reported to be the fun dormitory, but it was ragged and run-down. It didn't have all the amenities I thought he should have. USC was a big-time expensive private university, so it shouldn't be like that.

I said, "Let's go see the dean of students and talk about this." I said the dean, "I don't like the way this dorm is being maintained." He assured me repairs and upgrades would be done.

Wil asked me to talk outside for a few minutes. We excused ourselves. "I know your experience and your background, having been an RA in college and teaching at Howard," Wil said, "If you allow me to live in that dorm, I won't let you down."

"I'll take you at your word. Here's the criteria: If you earned all A's and B's with a 3.00 GPA or better, I'll buy you a brand-new car at the end of your first year."

He met my criteria. He didn't let me down. I warned him if his grades dropped below 3.0, I'll come get the car. He told me I could count on him. I dropped him off and went back to get processed out of Fort Ord.

I found out my name was not on the list for a senior service college. It was the first time I was in the window to be selected. It wasn't a big negative because I had only finished my command. I had been nicked for a second block on my first OER, which was the first time I'd been nicked in 15 years of OERs. Promotion in the combat arms was highly competitive, so I knew that had something to do with it.

The division commander called me in and asked me to be his division inspector general since I was not going to a service college. He wanted me to report directly to him, inspecting all units in that 15,000-man division. I asked for a couple of days to think about it. He told me to take as much time as I needed.

I decided to call Colin Powell because I considered him a friend. This was the first time I called him for advice. I explained the division commander's offer.

"That's all fine and dandy, but you weren't selected for senior service college," he said. "That's your first priority. Don't take that job. You get your family packed up and take them back to Washington and get on the Army staff. If you get another outstanding OER, you will be selected for senior service college and promotion to colonel."

I told my division commander that I would like to go back to Washington. He wished me luck and he hoped our paths would cross again.

We had accomplished so much in the battalion, but it was time to move on. One of my classmates from Florida A&M University, Lt. Col. Horace McCaskill, attended my change of command. When the command was given to pass in review, I marched by the assembled troops. The pomp and circumstance made me feel proud. I passed the unit's colors to my command sergeant major who passed them to the incoming commander. In the end, the troops passed in review. It was a happy and sad time. Command is what every officer in uniform strives for, to be the person in charge, whether it's a squad, platoon, company, or a battalion. Size makes no difference because you are the key person, the one responsible for all the unit does or fails to do. I didn't do it alone because I had outstanding officers, outstanding NCOs and outstanding soldiers and a wonderful wife and children who supported me.

In January 1983, the Army packed up our household goods and we packed up our car. First we'd stop in Los Angeles to see young Wil and say good bye. We arrived onto campus on early Monday morning with Lisa asleep in the back. I knew I wouldn't be able to take leaving Wil easily. I asked Emily to go in first, while I waited with Lisa. I stayed behind the steering wheel because we were parked illegally on this busy campus. I tried to rally my resolve to say goodbye to my son. The only thing I could say to him when I went upstairs was "promise me you will do your best." He promised and I gave him a hug. After leaving him, I started crying. I had to pull myself together in the hall before I got back into the driver's seat.

We traveled via the southern route back to Washington. I had my orders for the Army staff in the Pentagon, the Office of the Deputy Chief of Staff for Operations.

Our home in Burke Centre had been rented out when we left for California. The tenants' lease ended in August '83, so we lived in the Arlington area. We had never before lived in a predominantly Black community since we had been married. All my assignments living on post were integrated.

Lisa came home from school one day upset. She had a British accent, and she had never been around a group of predominantly

Black kids. On the middle school bus, a girl pulled her hair, taunted her, and said other nasty things.

That day, Lisa was waiting for me at the door. "Dad, I know what you told us, but do I have your permission to punch a girl on my school bus? I'm not going to fight her on the campus."

"Wait a minute. You can't do that. What happened?"

Lisa explained. I told her we weren't going to handle it that way. We were going to talk to the principal of her school about this girl. I called the principal and asked him to call the parents of the girl. I wanted a meeting with them and the counselor of the school.

Emily said, "I'm not going to get involved. You handle it."

For the meeting, I came in uniform from the Pentagon. At the principal's office he explained to the family why I was there.

"That's enough," I interrupted. "Let me take over." I looked at the father. I never looked at the young girl or the wife. Lisa was to my right. I said, "Sir, I've taught my children to never ever fight with anybody. I know brothers and sisters will push each other around, but your daughter yanked my daughter's hair and taunted her. Please tell your daughter to not do that again. Because if it happens again, sir, I'm coming looking for you."

"Wait a minute, are you threatening me?"

"Take it any way you want."

The counselor jumped in, "Please, Colonel Bryant, settle down."

"No, you settle down," I said to her and then to the father, "Do you understand me? You should know I'm a combat veteran and I know where you live." I didn't. "Tell your daughter don't do that, do you understand me?" He said he understood. I reached over and shook hands with him. "Thank you so much for taking time off from your job to attend this meeting. I have told my daughter to not even look at your daughter unless she absolutely has to. I'd like your daughter to do to the same thing. Lisa wanted to punch your daughter and I told her no. She won't, but if it happens again, remember me. Lisa, go back to class." I shook hands with the principal, and I said, "Have a nice day." I went back to the Pentagon.

On his spring break when Wil was visiting from college, Lisa asked her brother's help. She asked him to meet her school bus. He said, "No worries, I will pick you up from the bus." Wil was in great physical shape from all his athletics and he had on his USC football T-shirt on. He reported he jumped onto the steps of the bus as if to say, "Hello everyone." The commotion on the bus quieted down, but all he said was, "Oh! Lisa does have a big brother." She was never bothered again. Wil Jr. was always her protector. Lisa wanted that girl to see her big brother.

Chapter Thirteen

War College

My assignment at the Pentagon was a major cultural change from battalion commander to a general staff officer. With its 20,000 to 22,000 people working there, the Pentagon was a little city with its own medical facility, department store, bank, even a barbershop and many restaurants.

My new assignment on the Army staff was prestigious. Arriving, I thought I was terrific after my command, but the Pentagon has a way of deflating a person's ego. Lieutenant colonels, a nickel a dozen in the Pentagon, don't enjoy any special status there regardless of any previous assignments.

My introduction to life in the Pentagon was the first paper I prepared for my two-star general, and that could eventually be forwarded up the chain of command. I worked on it for two days, and it was two pages long. I turned it in bright and early in the morning. About noon, I received a call from my office secretary to come get my paper. I just knew it would be approved. Instead, it had a big X drawn through the center of the paper and "Start over" written in large print was across it. I was shocked at this development.

Fortunately, I had two friends in the office next door to me. One was my FAMU schoolmate and another officer I'd met previously. When I showed them the big X they volunteered to show me at the end of the day how to write in the "Pentagon style."

They teased me, "Unfortunately when you attended the Command General Staff College, you failed to take a course on how to write on the Army staff."

Pentagon writing was totally different. Verbosity wasn't acceptable. I quickly learned to be brief and to the point because the senior leadership didn't have the time to read more than a page and a half which had to include the task, discussion, and recommendation. My friends gave me all the "hot skinny" on that.

From then on, I never had a problem with papers. Everything was going smoothly until I was given an assignment to handle a special Army issue that the Army chief of staff, Gen. John Wickham, would personally make the decision to approve or not. As the author of the paper and a former battalion commander I would brief the chief of staff of the U.S. Army, which was mind-boggling, but I had confidence in my abilities to memorize and present clearly.

At this big presentation to the Army chief of staff, the deputy chief of staff for operations of the U.S. Army, a three-star general, and my boss twice removed from me, would be there, along with my two-star and the colonel I worked directly for. Additionally, 25 to 30 officers sat around the room as I was briefing the chief of staff.

The chief of staff entered through his private door and took a seat at the head of the table. Everybody stood at attention when he came in until he invited us to take our seats. Gen. Wickham had a lively sense of humor that helped put me at ease. I don't usually get nervous in briefing situations because I had thoroughly prepared on the topic and I always felt I knew more about it than the person listening. I had left the field, and I thought that I had more first-hand knowledge about the problem than the chief of staff.

Gen. Wickham announced the two issues here today and introduced me, saying, "Lieutenant Colonel Bryant, the floor is yours."

I looked him directly in the eyes as I discussed my initiative and I gave him my recommendation for the first issue. I presented the second issue, the same format, and gave him my second recommendation. I asked if there were any questions. He had several questions and I answered them. He went around the table to the three-star and the two star to see if they had any comments. They didn't have any.

"Bryant, you've done very well," Wickham said. "I think this is your first presentation at this level?" He was grinning. "I'm not going to approve your first recommendation, but I am going to approve your second recommendation, which is outstanding. Most people who come here to make presentations like you don't get anything approved, but you did get 50 percent. Good job for a first time. Is there anything else? No? Then carry on."

As I walked out, I realized my T-shirt was soaking wet from perspiration. I guess sweating in situations of extreme pressure like meeting Emily's great big dad and this briefing was me after all.

It was a humbling but inspiring experience for me. I felt if I could brief the Army chief of staff with his four stars on each shoulders, I, a lowly lieutenant colonel, could do anything.

On the home front, Emily was busy finding a job. Her ability to adapt was one of the many reasons why I was so successful because she supported me, and she could fit into any environment. My daughter was like her mom—she was very active in everything in school.

I quickly found I couldn't drive from our home in Burke Centre into the Pentagon every day because there was no parking for someone of my rank. Every morning I got up around 4 to 4:30 a.m. and was out of the door by 5 a.m. I went to the parking lot for the buses going to the Pentagon. I stepped into a slug line, a queue of people who wanted to catch a ride. With three or more people the driver could use the HOV lane and dodge traffic. I never thought Wilbert Bryant would hitch a ride as an officer, but I had no place to park. I'd driven to the Pentagon once or twice and parked in North Parking, which felt as far as Timbuktu. With the slug line ride, I would get to my office at 6 a.m. In the evening I was anxious to get home. I would leave my office around 6 to 6:30 p.m. and get home around 7:30 p.m. or 8 p.m., making for long days.

On those days, Emily was right there to take up the slack with Lisa. If Wil called, she'd handle it or have me call him back in the evening if he needed me to. I never arrived late to work for the whole time there because arriving late to work in the Army is blowing up your career.

One morning there was a message from my two-star written on the glass on top of my desk. I had been selected to promotion to full colonel, 0-6, and I had been selected to attend the National War College in Washington, D.C. Two thrills to start my day—I couldn't believe it. I knew that all my OERs had been outstanding except one. I had experienced how very competitive promotion was in combat arms, armor, artillery, and infantry, which is called the Queen of Battle. I was ecstatic to have been selected for colonel and the National War College because only a handful of officers were given that chance. Most officers will tell you about the three Senior Service Colleges, the Army War College, located in Carlisle Barracks in Pennsylvania, the Air War College is in Montgomery, Alabama and the Naval War College in Newport, Rhode Island. I was selected to attend the National War College which many consider the premier service college. There were about 170 to 180 officers in my class.

Among the class of officers and high-level civilians, one of my classmates was Francis Miko, my current neighbor. He came from a civilian agency. Also I made friends with many of my classmates. There were CIA, DIA, FBI, Secret Service—you name it, they had one of each.

My general army staff position hadn't lasted long. I came to DCSOPS in January 1983 and school was scheduled to start on August 7, 1984 with graduation in June 1985. The commandant of the National War College was U.S. Air Force Maj. Gen. Perry Smith. We were divided into about 10 sections with 17 to 18 students in each. Here I was rubbing elbows with some of the finest officers and high-level civilians in the federal government. I marveled how I got here. Sometimes even when you work hard, you don't get to that level of success.

The National War College was a challenging year of papers and presentations. The curriculum had a strong mixture of national security issues. We discussed the *Art of War* and the international security environment, the American policymaking process, and the proven military strategies. We could also choose elective classes and field studies. We also had what is called the Additional Voluntary Educational Opportunities program, trips for additional learning. A key trip early on was to the United Nations where we met with Ambassador Jeanne Kirkpatrick, President Ronald Reagan's UN ambassador. What an experience. The goal of the National War College was to prepare us for higher level positions in the federal government and the military. If you were a civilian, you'd be going up in the civilian chain. If military, it was for general flag officer positions.

I decided to take advantage of all I could while in the Additional Voluntary Educational Opportunities Program. I also took Red Flag Blue Flag, a simulation of war at an Air Force base in Las Vegas, but I passed on a visit to an aircraft carrier in the Atlantic. I'd had enough of rough seas already.

My major paper was on terrorism. Three other officers participated in writing that paper. Strangely as I reflect on our topic, I saw that as a future problem even then before 9-11. Having served overseas, I knew terrorism was going to be a major problem for the United States and our allies.

To complete my paper, I needed to get into the files of the FBI. I had called the research division, but I ran into a stone wall each time. I remembered my Florida A&M classmate, Big John Glover, who was the deputy director of the FBI. I told my teammates I'd call John Glover to get what we needed.

Everybody had a special section of the paper to complete. I called John and he said "give me a couple of hours and I'll get back to you." Not a couple of days or weeks—a couple of hours. He had a staff meeting coming up, but he said he'd take care of it.

About an hour and a half later, he called me back. Big John said, "All you have to do is call this number and someone will be waiting to show you around. They will have everything that you need."

My friend connection helped create my paper on terrorism. We didn't win an award, but we were recommended for a writing award.

While we were busy taking care of academic requirements, Emily was busy working, being a mother to our daughter Lisa, who was still in school, and she was also active with the wives of the National War College on their activities and excursions. On a noteworthy trip, the wives went to the Naval Observatory for tea to meet with the vice president's wife, Barbara Bush. Emily took the day off from work.

Of course she established a good rapport with the wives. Little did I know we would later cross the paths of former President George Herbert Walker Bush and Mrs. Barbara Bush, and also their son George W. Bush and his lovely wife Laura Bush.

The highlight of my second semester was an overseas trip. Students had the opportunity to travel all over the world. Countries were bunched, maybe seven or eight together depending on the area. I had always wanted to go to the Middle East because I had not been there in my Army travels, but the situation in the Middle East was very volatile at that time. I asked myself, did I really want to take a chance and visit the Middle East and be bombed by a country that didn't like the Israelis, or should I go where I didn't have to worry about that? I chose the trip to Africa.

Eleven of us traveled to three countries and spent six days in each. First we flew into Harare, Zimbabwe. Our very first trip on the first day we drove six to seven miles into the bush to a village to visit ordinary people. They were fascinated to see a Black American visiting them. We had a great time. They even played a joke on me. They said, "Wil, we want you to come up here." They tied me to a post! They stacked sticks around me like they were going to set a fire and burn me up. It was all a big joke. However, their poverty made the visit sad. A group of four or five officers passed the hat to take up a collection. Everybody put in some money, which we gave to the village chief.

Next, we drove east of Harare through a fascinating area near the Zambezi River and its game reserves. They put us on a Zimbabwean Air Force C-130 to fly us over Victoria Falls. I was on the left side of the aircraft. Plumes of water vapor rose hundreds of feet in the air around us. The plane bucked in the downdraft. As a qualified paratrooper, I was wishing we had parachutes. Next the pilot banked the plane to the left so everybody on the other side could see the waterfall. The pilot was fighting with the plane and the downdraft. Based on my experience flying in England, I knew how dangerous this downdraft was. As we were flying on a Zimbabwean aircraft, I didn't know and was concerned about the maintenance. But we made it.

Our next trip was Nairobi and Mombasa, Kenya. There was a Black American ambassador there, so I was chosen by our academic director to be the lead person to greet him. By the way, our academic director was also a Black guy, Lt. Col. William Isom, a smart military intelligence officer. I always wondered why he was not a colonel. He was knowledgeable about anything you wanted to ask about foreign policy and Africa. In Nairobi we visited both the ambassador's residence and his office. I briefed the ambassador on who we were, and why we were there, and what we hoped to learn on the trip. The embassy even had a very nice reception for us.

After three days in Nairobi, we traveled to Mombasa, where the U.S. had a port for the Navy to deliver supplies and equipment. There, they suggested we take a break and go into the Indian Ocean. We didn't have any women on the trip, just us and the professor, our academic leader. We donned our bathing suits and we walked out a half-mile into the Indian Ocean. The water only came up to our chests and we could see the bottom. So close to the equator, the water was hot like a Jacuzzi, but not so hot we couldn't enjoy it.

The final leg of the trip took us to the west coast of Africa. We flew into Cameroon for six days. We went to Yaounde and Douala. Cameroon was a former French colony and the official language was French. A representative of the government met us and began speaking in French to the class leader, our academic director, who answered him in French. I was surprised that this Black Army lieutenant colonel could speak French like a Frenchman. Emily had studied French, but her French didn't work in Paris because of her southern low-country accent. We had six great days in Cameroon.

Our trip had been a total of 18 days and it was time to go home. We discovered the majority of our trip mates, seven of them, had never visited Paris. We made a special plea to the academic director to contact Washington to extend our trip for two days, so we could go to Paris and spend a night and an afternoon to allow the guys to get a French meal and sightsee. I had been to Paris, but I voted yes. I had spent three or four days there during our assignment in Germany.

When we flew back to Washington, it was close to the end of our class. We turned our papers in, and it was time for us to get our orders for our next assignment. I had not received my orders before our trip abroad.

One additional fun event was the Little World Series. This is a game between the National War College and the Industrial College of the Armed Forces, which is located on the same campus as the National Defense University. The National War College had not won that game in years. With my high school experience, I decided to sign up to play. I also had several classmates who had played baseball in high school or college. This game was a softball game. We won the Little World Series on a grand slam by one of our classmates whom we called Boomer. He was a B-52 fighter pilot, Col. Kenneth McAlear.

We came off that cloud and got serious about getting our next assignment. I had good friends in that class, Col. Eugene Russell and Col. William Johnson. In every seminar the three of us would sit together when it was held in the main Assembly Hall. I had met Gene Russell back in May 1980 in a pre-command course for future battalion commanders. I ran into him again in the Pentagon during my DCSOPS assignment and he was in my section at the National War College. We are friends to this day and with his wonderful wife Sharon. Other good friends were Col. Jim and Sissy Williford. Their daughter and our daughter both attended Princeton University. Later on, both girls were on the Princeton cheering squad.

When I hadn't received my new assignment, I decided I'd use my friendship with Gen. Colin Powell again. By this time, he was senior military assistant to the secretary of defense, Caspar Weinberger. I told him I needed an assignment. I was on the alternate list for brigade command, but I hadn't been selected. In my career field, if I didn't command at the colonel level, then I wasn't going to make general officer. It was rare for alternates to be plucked off the alternate list for command. Speaking to Gen. Powell confidentially, I asked if he could make a recommendation as to where I might try to be assigned. He said he'd get back to me in a few days. Sure enough, after a couple of days, he called back to put me in touch with one of his senior assistants. I received a call from his assistant who asked me to come over to the Pentagon so he could show me around as a recent War College graduate.

I had been promoted to colonel in January 1985. Emily pinned one of my Eagles on and Maj. Gen. Smith pinned the other one on. I was looking for a position that would use my promotion, background and experience. Nearly as bad as majors, colonels were 10 cents a dozen in the Pentagon.

Gen. Powell's assistant discretely showed me two positions. Both were commensurate with my schooling, but I didn't want to be in the Pentagon's basement, with no windows. The rats down there were as big as I was. There were many stories circulated about rodents in the Pentagon. After the tour I thanked him for spending his time showing me around. I had been down in the basement on previous assignments. I told him I'd get back to him and I'll let him know of my decision.

I decided I was going to find my own job. I called military personnel and told them I wanted to compete for a particular position. The assignment officer agreed I was well qualified for it. I put the phone down and called Gen. Powell and thanked him for setting up my tour of Pentagon positions, but I thought I had another position and I'd let him know. He said good luck and he'd be in touch. As a result of my efforts, I was selected to return to ODCSOPS as chief of the Joint Command Post Exercise Division in the Pentagon.

We were rapidly approaching my graduation day. On this big day, in the audience was Emily, my daughter Lisa, and my brother Sylvester Bobby Bryant who came down from New Jersey. In my NWC yearbook they captured those three sitting in the audience in a beautiful photo. We walked up, saluted, and got our degrees.

My next assignment was another Pentagon assignment, one more time. I had a team of officers and civilians in my division. Maj. Gen. William Moore was my senior rater. I had a brigadier general as my immediate rater. I handled a variety of training initiatives for the Army, a very important job.

Lisa had told me early on, "Dad, if the Army decides to move you again, Mom and I are going to stay in Washington because it's time for Mom to put her roots down and work in her area of expertise."

In view of Lisa's statement, I did everything I could to stay in the Washington metropolitan area. I figured I would probably retire in this area.

In the Pentagon, a new boss, Maj. Gen. J. D. Smith replaced Gen. Moore, who had been a fantastic boss. We worked hard for him; he rewarded me with an outstanding OER. Same thing for Gen. Smith. Internally I knew I had to get out of the U.S. Army because I had not been selected for 0-6 command. Even though I had graduated from the National War College, I knew if I had not been selected for 0-6 command, I had only a 50-50 chance of making brigadier general. With this thinking, I was ready to return to academia because of all it had done for me and for my family. I also wanted to give something back to society. Plus, I wanted to honor Lisa's request to stay in Washington and give Emily an opportunity to work in her area of expertise. I decided to move out of this Pentagon position and find another job.

A while back, a friend of mine had told me about the possibility of getting what is called a "Black Book job." There was actually a black-bound book sitting in a strategic office in the Pentagon. It listed all the available jobs open at the very highest levels for generals and colonels. I found the book, but I didn't tell anybody. I researched the information on the position that was a senior military assistant to one of the deputy assistant secretaries of defense for reserve affairs. I talked to the officer assignment division and told the assignment officer who handled my assignments I wanted to compete for this position. I wanted to move into the office of the Secretary of Defense. The assignment officer told me I was highly qualified and he'd add my name to the list for consideration. He informed me there were five other officers being considered for that position. Three of those officers had commanded at the colonel level where I was aspiring to go to. I told the assignment officer I'd take my chances.

I received a call setting up an appointment for my interview with the deputy assistant secretary, the key person himself. When I reported to him, he asked me an assortment of questions. I looked him in the eye and confidently answered. I thought I handled myself well because I had performed well in so many different jobs in the Army at that point in my career.

Two days later I received a call from the deputy assistant secretary, Brig. Gen. Carl Morin.

"Colonel Bryant, I'd like for you to come over. I have selected you to be my senior military assistant."

I said, "That's great news."

When he asked if I had any questions, I said, "I have one question. I know there were three other officers who were competing for this job. Why did you select me?"

"One key reason. You have a strong background and experiences that none of the others had. You have a good knowledge of the planning, programming, and budgeting systems of the Pentagon. I need someone who can hit the ground running when they arrive for work on day one. I don't have time for an officer to come in and have to be trained to operate those complex systems. You checked all the boxes. Congratulations. When can I expect you to report?"

I said I'd have to give my boss two weeks' notice, and I'd join as a senior military assistant.

Chapter Fourteen

High School Reunions and High School Star

While I was busy in the office of the Secretary of Defense, surprisingly, high school both for me and Lisa became important.

Supposedly only the very finest officers were assigned in those key positions, especially as a senior military assistant to one of the deputy assistant secretaries of defense. My boss, Brig. Gen. Carl Morin, was the direct but soft-spoken officer who interviewed and selected me. Although we were close in age, he had received a "below-the-zone" promotion to propel him to brigadier general while I was still a colonel. That's a common practice found in all services, not just the Army.

In our unit, we had five major divisions; this was a joint assignment with all the services, Army, Marines, Air Force, Navy and Coast Guard. The five divisions were headed up by colonels, which makes it difficult for another colonel to be a notch above of them as a senior military advisor. I was not in their direct rating chain, but I had total responsibility for all of the administrative papers coming from those five divisions. I had to check all documents before they went into the deputy assistant secretary. It might appear to be an easy job because the work came from colonels, but there was always tension, especially with one.

Each morning when I was in this job, I would leave the office and go to the Pentagon officers athletic club, for two hours of exercise. I believed if a person did not keep physically fit he would more than likely end up having medical issues like a heart attack because of the stress. I would often see a golf cart used as an ambulance in the corridors on its way to the Pentagon's medical center. I didn't want that to happen to me.

One morning when I hadn't been on my job very long, I returned from the club at 11:30 and greeted the three administrative assistants who reported directly to me and one to the deputy secretary. My usual greeting was, "Hello ladies, anything going on? Everything quiet?"

As soon as I said that, I heard my boss, the assistant secretary of defense, call, "Wil, get in here."

"Good morning, Sir. What's going on?"

He threw a decision paper across the desk at me. "There are two mistakes on this paper."

I grabbed it and said, "I have no knowledge of this. Any papers that come through me, my initials are in the upper right corner of the first page, which means that I have seen it and take full responsibility for the content and any typos that may appear on it." I examined the paper closely and I knew where the problem was.

A Navy captain in the back didn't like the idea of sending his papers through me, but I was the senior colonel assigned and I had big time date of rank on him. Plus, I was a graduate of the National War College, which he was not. I told the general I'd take care of it.

Fuming, I went to the captain's office and slammed the door. "Mike," I said, "You bypassed me with this paper. It has two typos and that is totally unacceptable. Let me tell you one thing, and I'm only going to say it once. If this happens again your rear end is going to be like grass and I'm going to be the lawn mower. I'm going to tear your rear up and you will not be there."

"Wil," he said "I don't work for you."

"I know you don't work for me, but all your administrative papers go through me to the general. If you do it again, you are going to get your rear end fired." I used some expletives in my language. Of course, I meant re-assigned. With my "military hat" I got him squared away. I said, "Do you understand me?" He said nothing. I slammed the door again on my way out. Several pictures fell off the wall.

I went straight to the general to tell him I had spoken to the captain, straightened it out and explained the consequences if our administrative procedures were not followed. It never happened again. Everybody got the message. We moved on and we worked in peace.

During that time, a lot was happening in my family. First of all, my high school baseball coach, Earl Dinkins, contacted me. When I was playing baseball for him, he always called me 'Son.' "Son, I want to ask a favor of you."

"Coach, anything for you." I still called him coach, even though high school was 30 years ago. He wanted me to come to Florida and organize an all-class reunion for our high school.

Coach Dinkins continued, "Our high school existed from 1954 to 1969. When integration finally came in 1969, our high school was downgraded to a middle school. I want you to take the leadership for this reunion."

I said, "Coach, I love you but there are hundreds upon hundreds of alumni in the Miami-Fort Lauderdale area who can do this task. I'm in the Pentagon in a tough job."

"I know where you are. You and your family have been outstanding. You all have excelled in all areas, academically, athletically, and professionally. You managed troops, you fought in Vietnam, you commanded a battalion." He certainly had done his homework. "You're now a colonel. There are a lot of personal animosities among the folks here, and they won't listen to anyone else. You are the ideal person to get this done."

I finally agreed. Saying yes was my way of saying *Thank you, Coach, for giving me the chance to play on our championship team.*

Immediately I called a meeting of students I knew both in my class and the classes before and after me, about eight or nine individuals. When I told them the purpose of the meeting, my friends told me they had attempted to do this before but they had confidence I could do it and offered their support.

I called the Marriott Hotel and booked a conference room set up to accommodate 35 people. I asked for notepads and pencils, a setup of hors d'oeuvres, fresh vegetables, and bottled water. When I flew in for the meeting, 35 to 40 people didn't show up but close to 100. We asked hotel management open some curtains and set up more tables. I was elected as first president of the Mays' High School National Alumni Association, which gave me the leadership position to get us organized and accomplish the reunion on time. I held meetings twice a month, flying from Washington to Miami. I defrayed all my costs associated with it primarily because I believed the reunion was an important task. Although I was spending my own money, I had Emily's blessing.

Coach Dinkins also wanted us to preserve the school's history, honor the esprit de corps, and amplify the respect we had for the teachers and the staff of the school. He didn't want our high school history to disappear. Our meetings energized the alumni. I formed nine different committees to get us organized for the All Classes Reunion. The alumni group worked closely together to honor our mutual past. Coach Dinkins knew my two older brothers would likely chip in and help. They gave me some great advice, but I was the one Coach Dinkins held responsible for the first reunion for 1989.

Our group decided on a four-day event including a mixer, a dinner dance, a picnic for all classes, 1954 to 1969, and a special baccalaureate program in our old school building. I arranged the dinner dance to be at the officers club at Homestead Air Force Base. With my "colonel hat" on, I asked the manager for the use of the officers club, the dining area and the dance area. For our final event, we asked and were given permission to use the school's auditorium so all the alumni and all the teachers could attend.

At that same time, Lisa had started high school. She loved her brother as her hero and she thought he could do anything. She'd always call on him when she had a question.

When she entered Robinson Secondary School as a freshman, one of the first events her first year was the contest for Miss Robinson Secondary School. Here Lisa was, a young Black female at a high school of over 3,000 kids in Fairfax County, one of the wealthiest counties, not only in Virginia but in the United States. Lisa decided to compete for the Miss Robinson title. We had always told her you can do just about anything you want to, all you have to do is try and give it your best. She was a pretty girl, but more importantly she was smart and well spoken.

In this contest, 70 to 80 girls participated. The first round of competition tested talent, and they narrowed down the number of contestants to 10 students. Lisa was named among the top 10. For her talent performance, she played the piano beautifully. We were thrilled for her. The next cut was time to name the top five contestants for Miss Robinson. The final round was held in the auditorium which was filled with 300 to 400 people. Her school was so big they even had two or three sub-school principals under the lead principal of the entire school. Among the Top 5, they called her name, Lisa Nicole Bryant. The place erupted with applause.

I whispered to Emily, "They selected Lisa because she is only Black girl in the top 10.

"That's probably true," Emily whispered back. "Let's see what happens."

To make the final selection, those five young ladies faced a series of questions. The audience was riveted. The girls had to respond extemporaneously and give a reply how they would deal with a particular issue. When it was Lisa's turn, she confidently addressed the question. She did an outstanding job; of course most parents would say that. Lisa knew if she flubbed, her dad and mom would tell her, but with her performance, we certainly didn't have to.

The time had come for the judges to decide the winner. They announced the third runner up. It wasn't Lisa. Then the second runner up. Next was the crowning of Miss Robinson Secondary School. They announced, "Our winner is Lisa Nicole Bryant." All of a sudden, there was an eruption of applause for Lisa. She was selected as Miss Robinson Secondary as a freshman at this huge 3,100+ high school. It was a major accomplishment for her, all based on her scholarship, beauty, poise, talent, and ability to speak. We were thrilled for her and so proud.

Lisa was also elected captain of the school's cheering squad. I always wondered how I would deal with Lisa's events because I was accustomed to Wil playing sports. I soon learned I loved watching hers, too. For her to be voted captain by her peers meant she had excelled in many areas, including strong leadership qualities.

We always attended her games. She would always come to join us in the stands at halftime. She'd ask me how was she doing. I'd tell her she was doing great. I didn't pay attention to the game, I was too busy watching the squad cheer.

Lisa also became the president of an organization, the Junior Civitans Club. She was also selected to participate in Kids to Kids, a program established for kids in the United States to talk to kids in Russia. The U.S. kids asked the Russian kids questions about their lifestyle, their school, dating or any topic of interest. This program went on for about two hours. All the newspapers covered it. Again she really excelled. Out in California Wil and his girlfriend listened on the radio. Even the girl's grandfather heard it and was so impressed with the little lady he had met only a few years earlier as a 12-year-old when she visited Wil Jr. in California.

After a few weeks at Robinson Secondary School, we attended a Back to School Night. When the general orientation ended, parents would break out and go to different classes. We decided to visit Lisa's English class to meet the teacher. We knew English was so important for a child in the school system and in college and then professionally. If a person cannot effectively communicate orally or in writing, she or he would not be a success. We went to the class, sitting in the students' seats. Usually the teacher would explain to the parents what the syllabus entailed.

In this class, the teacher said, "Rather than me give you all a thumbnail sketch of what this year is like, I asked one of my top students to explain it to you. That student is Lisa Bryant."

We had no knowledge she was going to present. Lisa spoke courteously to the parents and gave a very thorough overview of what they were expected to accomplish that year. She used no notes and had memorized the major points, just like her dad. She asked, "Are there any questions?" She said she'd take their questions and answer if she could or her teacher would. Some of the parents raised their hands and Lisa responded. The teacher would jump in every now and then to add something. Lisa did this for about an hour. I thought, *My oh my, was there anything our Lisa can't do?* Our kids never failed to surprise and impress us with their accomplishments.

While we were stationed at Fort Ord, California, Wil had been selected from Seaside High School to attend Boys' State in Sacramento, where he had a great time. Lisa was selected to attend Girls' State in the Commonwealth of Virginia at Longwood College in Farmville, one of Virginia's 16 public colleges and universities. She was among a group of 600 to 700 students. She loved meeting people and participating in activities like that. She considered herself a military brat who had traveled, had seen and done so much.

Lisa even competed to be the governor of Girls State. I didn't think she'd win, but the experience of trying her very best and competing to be the governor would set her ahead of her peers. She recruited her mom to make flyers to pin up throughout the campus for her candidacy. Election day came, the vote was taken, the leader of Girls' State came to the podium and announced "the new governor of Virginia is Lisa Nicole Bryant." I couldn't believe it. Out of all those girls! Again it was the way she handled herself, extemporaneously like her old dad at his briefings. Confident and articulate, and with a strong campaign of her platform: all that sold her to the other girls. We were very proud of her.

Just as I had promised Wil, I told Lisa if she earned a four-year ROTC scholarship and maintained a B average during her first year of college, I would buy her a brand-new car. Lisa also earned a four-year ROTC scholarship that paid all her tuitions, fees, books. It made me so pleased when she was notified of it.

She said to me, "Dad I'm going to be just like you. I think I'm going to surpass you."

"Sure, let's try it. The game is on."

She was told about the scholarship in November her senior year. Since I knew Lisa would maintain a 3.0 GPA. I put plans in motion to buy her a new car before she graduated high school. The car would make everybody's lives easier. We had moved from the Burke Centre school district into another school district. But Lisa wanted to finish at Robinson Secondary rather than change schools in her senior year, which meant Emily took Lisa to school every morning and back. I said it was okay as long as we could accommodate her as a family. Emily worked, and she provided Lisa's transportation; she was a super mom. I didn't think Lisa would let me down if I bought her car early.

I made a plan to surprise her. I went to Hershey, Pennsylvania, to buy a new car for her, just like I had done with Wil's car. The reason I chose Hershey was they didn't sell many Mazda models up there in that small town. If I bought a car there, I could save a little over $2,000 on a new car.

One night, Lisa had a game, cheering at the school's football game. I bought a one-way ticket on the Greyhound bus to Hershey. I had already coordinated with the Hershey's car salesman who remembered me from when I bought Wil's car. I told him I would be there early because I wanted to get back to my daughter's football game. I picked up the car, a brand new 1989 Mazda MX6. I told Wil Lisa wanted a different car from her brother's, if I bought her a car. I drove it fast to get back before halftime. I joined Emily in the stands.

Lisa, as usual, came up into the stands. She gave us each a hug. I told her I had a surprise for her.

I said, "I have a brand-new car out in the parking lot for you."

She said, "Oh Dad! Let's get it!"

I said, "Oh no, you have to continue cheering for the full game."

She said, "I'll call my co-captain in and have her finish for me."

I said, "No, no, no. It's dark. I'll drive the car home and then I will turn it over to you." She was so excited and I was excited for her because she had earned the right to get her new car. The first person she called was her brother to let him know Dad had bought her a car. Here the first semester of her senior year hadn't even ended. Wil had to wait until his freshman year in college. He asked me why did I buy Lisa a car now and why didn't I do it like that for him? I told him the situation and the times were different with Lisa, which is why she received her car early.

At home, I gave Lisa guidance about her car. She was to be the only person to drive it. When she went away to college, she could only drive on weekends. No more than three kids could be in the car. No one could drink or smoke in the car. Those were my rules for high school and for when she went away to college.

Many universities recruited her because she was a National Merit finalist. Duke University even sent one of their alumni in the

Washington area to take her out to lunch and sell her on Duke. Lisa had other plans. My mother lived in Trenton, New Jersey. Lisa loved my mother not only because she was disabled, but also because my mother was a visionary. She had purchased Lisa's piano. Lisa wanted to be close to her grandmother geographically, which meant she was going to accept enrollment at Princeton University. She'd be only 15 miles away from her grandmother, so she could pick her up and bring her to campus so she could see whatever Lisa was doing, cheering, playing the piano or whatever. That's the way Lisa was.

During her senior year, one night when I came home from the Pentagon and my car was in the shop overnight, I told Lisa I would like to use her car to drive to Park-and-Ride and use the slug line to travel to work.

She said, "Oh no, Dad I'll get up and take you." I reminded her I got up at 4:30 a.m. and was walking out the door by 5. Lisa continued, "You told me you didn't want anybody to drive my car and that includes you!" I knew my guidance had sunk in; she was adhering to my advice, literally.

Chapter Fifteen

Shaking up Virginia Union University and Virginia Republicans

My job in the Office of the Secretary of Defense kept me busy dealing with those five officers, but I kept it a smooth operation. On the home front, Emily had stayed in touch with our friends Tony and Marcia Manning. Tony had retired from the Marine Corps, and he was working in Richmond at Virginia Union University. Marcia told Emily that Tony was aware of a vacancy at the university and that I should give Tony a call. When I did, he said the president, S. Dallas Simmons, was looking for a person to fill a high-level position. Tony suggested I interview and get practice in the job interview process in the civilian world. I hadn't even written a resume or anything. He told me to put one together and send it to him to give to the president.

I was invited for a secret meeting with the president on January 3, 1990. When we started talking, it was clear he knew quite a bit about me, probably through Tony.

Thirty minutes into our talk, he said, "Colonel Bryant, I'd like to offer you this position."

"Hold up, Mr. President. We haven't talked about a specific position." He explained the position was vice president for student affairs. I mentioned I didn't have a Ph.D. or an Ed.D., only a Master of Education, with a concentration on Student Personnel Administration. He told me I didn't need another degree with all my experience in the Army and at Howard University. He was certain I'd done more than most vice presidents. He said he'd want me to start July 1, 1990, six months from now.

"Oh, my goodness. I don't make unilateral decisions about my career. I always consult with my wife." He offered me a cell phone and asked if would I call her right then.

I called Emily and told her Dr. Simmons had offered me a position. Emily asked what were we going to do if I got out of the Army. I surprised her with "we are going to be a university vice president."

"Oh!" she asked, "What is the compensation package?"

I didn't know, but I did know what my retirement pay from the Army would be. Me—the math major—I'm standing in the hall doing the numbers in my head. This opportunity would be the ideal time to exit the Army because I had made the alternate list for 0-6 Command twice, but was not selected.

I was also ready to give back to academia because it was responsible for my success. I told Emily I'd accept the position. She agreed. Back in his office, I told him he had a deal. We shook hands on it. He reminded me again to keep the appointment secret until his press conference in early June 1990, when he'd announce me and several other key positions. He asked me to walk the campus with Mr. Walter Miller, the vice president of university services for a private tour to show me my areas of responsibility, Residence Life, Student Life, Student Health and Career Planning and Placement. I had a good working idea about student affairs from my Florida A&M registrar's office student worker experience. Dr. Simmons had even heard about that, too. Mr. Miller and I walked around, giving the impression to all the students and instructors we were only conversing. Back to the president's office, he said he looked forward to seeing me at the press conference in June 1990.

At the Pentagon I went to see my two-star boss and gave him the news. He wasn't very happy about losing me. I told him it was time for me to look toward my second career. I gave him notice I was retiring from the United States Army June 30, 1990 and accepting this new university position in Richmond, Virginia effective July 1, 1990. He was glad I gave him plenty of time to find a good replacement for me.

My June 30 retirement date came fast. First there was a ceremony in the Pentagon for me to be officially retired. At Fort Myer, Virginia, all the officers and high-level NCOs who were retiring in June 1990 were honored on that ceremonial field. Here I was, almost 28 years later saying good-bye to the Army. I knew that I was going to miss my friends, but I wondered if I was going to miss the Army and those wonderful milestones. Still, when I make a decision, I don't look back. The ceremony was quite a colorful fantastic day.

Emily put the final touches on my retirement. She threw a retirement party at our home at Fairfax Station and invited many of my friends. Our two children joined us. She had a three-piece band upstairs on the landing with a dance floor on the main level.

When the day ended, I took off my uniform and gathered my suit and tie. The next morning, I headed to Richmond, Virginia. Earlier I had called my executive secretary Sandra Burrell and asked her to alert all my division chiefs that I wanted a meeting with them at 10 a.m. on July 1. Just like with my officers, I wanted them to meet me so they understood my goals and objectives and to clarify what I expected of each one of them individually and collectively as a group.

On the day of the meeting I arrived around 7 o'clock, three hours before the office was open. I had to ask Campus Police to open it for me. I'm a hero one day and the next day I'm vice president of a university. At 9:55, I called Ms. Burrell aside to ask if we had a full accounting of everybody. She said one person, the director of residence life, was not there yet. We waited until 10 a.m., and I closed the door and began the meeting. About 10:20 I heard a knock on my office door; it was that other director. He explained he had been on a personal phone call in his office which was nearby. I knew where his office was, not far from my outer office. All I did was say hello and welcome him to the meeting.

When the meeting ended, I asked this director to wait. "You came to my meeting 20 minutes late. I had announced this meeting 30 days in advance. Sir, I'd like you to turn in your keys and all the papers you have from your office. I'm terminating you today at 5 p.m."

"Wait a minute, Mr. Bryant, you can't do that."

"What do you mean by that? You don't have tenure at this university. I don't either. This new contract started today. I don't tolerate anybody coming late to my meetings. So do as I asked you to do."

"This is not the end of it," he said. "You'll hear from my lawyer."

"I'll be happy to talk to your lawyer. Be out by 5 o'clock." I was prepared to have the locks changed to his office.

He kicked up a fuss. He went to see the university counsel and the president. They told him Mr. Bryant was the one in charge of that office and that he was the person who signed off on all contracts for all the directors. He was told that whatever legal action he wanted to take, go ahead and do it. I'm still waiting for him and his lawyer to show up.

With only a few bumps, we were off to a good start. I needed to learn the general operation of the university fast. I wanted to establish a good rapport with the other cabinet members, especially the provost and the vice president of academic affairs, Dr. Gerald Foster. Another key person was the vice president of financial affairs because all my money came from him to support my programs. And also important was the vice president of university services, Mr. Walter Miller. His crews would be cleaning all my dormitories and all the buildings we occupied. He also operated the dining facility, which also came under my area of responsibility. The vice president for university development was my friend Col. Tony Manning. In fact

I resided with him for my first two years until I found a place to live of my own.

The final person to get to know was the president of the university, his idiosyncrasies, and his first lady. I had to understand the campus politics because, in the military, we operated one way; on a university campus, I knew it was totally different. In the Army when I gave an order for something to be done, it was done. In the university environment, I understood it didn't happen that way.

The first two or three weeks after the students began classes, my routine, after my regular 8 a.m. meeting with my immediate staff, was to visit the residence halls to check their cleanliness and the state of repair. I was responsible for the welfare and the safety of the students so I wanted to insure everything worked properly. Parents were spending a tremendous amount of money to send their children to this university; I wanted to make sure they were getting their money's worth.

After two weeks, I sent a written report to the vice president of university services detailing the deficiencies I had found. Mr. Miller didn't like that. He went to the president and told him Col. Bryant was inspecting the dormitories and facilities as if he was in the military, with a white glove inspection. Even though he ratted on me, it didn't bother me. The president called and asked me to talk with Mr. Miller. I asked Dr. Simmons if he was dissatisfied with what I'm doing. He said, oh no, the opposite—he was very happy. I was doing exactly what he wanted. He received many complaints from parents about things not being operational like closets, sinks, and restrooms, not being clean.

I invited Mr. Miller to my office. I said, "I want to work in peace. We have a job to do. We are here for one reason—the students. In the area of Student Affairs we assist the academic process. Students want an education to improve themselves so when they walk out of these hallowed halls, they will be prepared. We must work on the same page about maintenance and safety."

Walter Miller thought it over and agreed. He became one of my very best friends during my VUU time. When I was out inspecting daily, he'd join me many mornings.

One of my first challenges was to reestablish the esprit de corps, the morale among the student body. In previous years, they did not even have a homecoming celebration. No homecoming parade! I told the president of my intention to hold a homecoming parade in late September or October and have a full homecoming week. He really appreciated my drive, energy, and thoughtfulness, but usually he told me when they had a homecoming, they started planning in January. He didn't think I could pull it together.

"Mr. President, can't is not in my lexicon," I said. "As long as I have the money on my budget line, all I want is a green light to go ahead and get it done." He told me to go ahead and see what I could do.

I immediately formed a homecoming committee of 8 to 10 people from several offices on campus and I recruited a lady willing to help me. I even invited professors for their input.

The key obstacle was permits. The president was right in one respect; you start in January because Richmond granted permits for parade routes and this took time. My preferred route was West Broad Street to Lombardy Street, passing in front of the university. The city denied that route so we planned an alternative. We had about 50 parade units in this first homecoming parade, a very low number by my standards. I had wanted close to 100 units, but it was a start, the catalyst to bigger homecoming celebrations.

The second major challenge was to get the Student Government Association involved in other activities on the campus. For example, they didn't have an annual yearbook. One hadn't been published in several years. I believed that was a no-no. Parents and their students want some record of their college years. I immediately appointed a student to be the yearbook editor. I told the vice president of financial affairs I was going to move money around for the yearbook to be published. I had already contacted a publishing company and I was determined to publish a yearbook.

The financial affairs vice president, Mr. Nathaniel Lipscomb, wanted to know if I had discussed this idea with Dr. Simmons. I told him I had, and he had approved. He wanted to check with the president. I encouraged him to check with anybody he wanted to, but we're going to do it. Lipscomb put up a lot of resistance, but when he saw that I wouldn't take no for an answer, he fulfilled my requests. VUU also didn't have a student newspaper. I told the university leadership we needed to publish a student newspaper at least once a month. I gained approval and this task was accomplished.

The last piece of the school spirit puzzle was the student leaders. The Pan-Hellenic Council played a big role on the campus, but they needed to raise their profiles. I called a meeting with the Pan Hellenic Council and gave them guidance on how they should function on campus, to include a no-hazing policy, and following the university's rules. I advised them if they violated the rules, suspension would occur. Depending on the severity of the violation, a Greek organization could possibly be kicked off the campus. The word of my policy spread quickly on campus. No one had enforced this policy before.

The next person I wanted to build rapport with was the Richmond chief of police. At that time, the drug war was very bad in Richmond. I made it my personal business to talk with the freshman class about life off the campus. I instructed them how they should conduct themselves when they left the campus and what I expected of them on the campus. Getting in fights wouldn't be tolerated. If a fight happened, I would find out who started it and suspend that student. Unfortunately, I had to make an example of seven or eight students to get their attention. I received quite a few calls from parents, but it didn't bother me. They even came in to meet with me. The president supported my policies because I was trying to create an environment of learning. The students shouldn't have to worry about any distractions on campus interfering with their education.

If a student was dressed inappropriately, especially some of the boys, I would call that young man to my office and talk to him personally. In most cases, I'd tell him to go back to the dormitory and change his clothes. I would let the student know I didn't like it and for him never to wear inappropriate clothing again. Otherwise, I would suspend him. I had to do that to two students. In each case their parents came out to see me, and I explained what happened. I told them I'd see their student the next semester. After I had been there six months, the students knew that they had a person interested in their well-being who wanted them to have a great experience while simultaneously getting a good education.

While on campus, I looked at current politics in America because I've always been interested. After being in Richmond for several weeks, I called the State Democratic Party and asked them for information about their organization and their current initiatives. I grew up in a family that voted Democrat. My first vote for president was for President Kennedy, when I was starting on active duty. I was really hurt when he was assassinated. In Richmond, over two weeks rolled by, and I didn't hear from Democrats.

The next week I called the Richmond Republican City Committee and made the same inquiry. Not only did they show interest, but I guess I shocked them. Here was a retired Army colonel from Virginia Union University asking for information. I was invited to their headquarters that same day. I spent a couple of hours with a young guy, Kirk Schroeder. I said I was interested in joining the city committee. I had no motivation to occupy a position or office, but I wanted to get involved because I always remember the old saying "for whom much is given, much is required." This activity was another way I could give back to others. With my almost 28 years in the military, and traveling all around the world, working with all kinds of people, I thought I could make a difference. They told me they were delighted to meet me. They asked me would I like to wait while they secured me a membership card from the city clerk's office. I was made a member of the Richmond Republican City Committee on day one.

At the first city committee meeting I went to, the organizers introduced all the new people. I glanced around this crowded room and was surprised there were only two Black people in this meeting. In my 28 years in the military, I felt color blind, but it was strange. They introduced me and asked me to tell them about myself. I mentioned my childhood in South Florida, my education at Florida A&M University, and my Army career. Now as a vice president for student affairs, I looked forward to working with them to help people in the Commonwealth of Virginia.

At VUU, a big part of university life was sports. Football, basketball, women's basketball—everything to do with sports. I devised a strategy to keep the players from having problems with me and the administration. Football players can be kind of rowdy. I called for a meeting with head coach Joe Taylor. He needed my support, and I needed his. I asked to meet with the football team co-captains and the athletic director to discuss what I expected of them. We wouldn't tolerate any wild or silly behavior, but I wanted them to have fun. The head coach really liked my approach. Again, I heard "no one has ever done this before." I told them about my experience at Florida A&M University, one of the finest HBCUs in the United States. They were in awe of the glory days of FAMU. I enjoyed developing a good rapport with them.

I repeated that strategy with the head basketball coach, Dave Robbins. He had a championship-caliber basketball team; we had no problems. The women's basketball team was also very good. I met the head coach, all key players, captain and co-captain, and some of the trainers and told them what I expected. They were leaders on the campus, and I expected their conduct to be exemplary, setting a good example for the general student body. I warned the team leaders, if they got rowdy, carrying on like they were better than the others, we would have to have to shut that behavior down.

Next the university was a participant in a program named Carver Promise. The president assigned me as the representative from Virginia Union University. This program included area universities, the Virginia Commonwealth University, the University of Richmond, Jay Sargent Reynolds Community College and Virginia Union University. The college students from each school mentored elementary school students in the local community. The program started with third graders. Each one of higher education institutions was required to provide college students to assist at their selected elementary school. When I attended the first meeting of the Carver Promise, only four VUU students were there, which I found totally unsatisfactory. When I returned to Virginia Union, I changed that equation. The next meeting we had over 20 students from VUU participating to get us on par with the other universities.

In the program, we'd stay with that elementary school class all the way through high school. We also contacted the corporate world in Richmond for donations to help the elementary students. We organized educational activities, field trips and museum visits to whet their appetite for education.

The Carver Promise gave me an opportunity to meet my counterparts, the vice presidents at the other universities and the president of the community college. I learned about their student activity programs, knowing the public institutions supported by Virginia taxpayers had a lot more money than Virginia Union, a private institution. Some programs they offered we could implement. For example, we added movies on Friday nights to give our students another option for fun and to keep them on campus.

At Virginia Union University, the basketball team won the Division II Basketball Championship. The football team won in their division and were Central Intercollegiate Athletic Association champions. I knew all the stars on those teams and I helped to keep them grounded. I was tickled that most of them called me Colonel Bryant.

Our football games were a big deal. The president also gave us, all his cabinet members, the opportunity to sit in his box at our football games. I did that initially, but I decided I didn't want to be around the president at the game. I really got into games, cheering noisily. My preferred seatmate was vice president of university services, Walter Miller.

I established a Student Leaders Program that included some of the finest students on campus. I brought together about 15 students to work at convocations, at football games, or to escort VIP visitors on the campus. These students had good GPAs and they carried themselves with dignity and pride. I set the organization up, and it further enhanced the student morale and esprit de corps on campus.

Lisa visited Virginia Union University to have an informal get-together with the cheering squad. Needless to say, I was proud of her. News spread quickly, "Colonel Bryant's daughter would be visiting Virginia Union University today." I told her to be on her best behavior. She laughed, saying, "Dad, don't tell me that. You know I will."

For my second year we began planning our next homecoming parade in January. I visited the Capitol and met Gov. Doug Wilder, who had just been elected, the first Black governor since Reconstruction. Wilder was a graduate of Virginia Union University. He wanted to know a little bit about me. I wanted him to be the Grand Marshall for VUU's second homecoming parade. He said he'd love to do that. I then went to city hall to see the chief of police to invite him to be in the homecoming parade. He readily agreed. My strategy to use West Broad Street Route as the main parade route was approved by the chief of police. The homecoming committee for our second homecoming parade was delighted. We had almost a hundred units participating. I was able to get a Color Guard from Fort Eustis and another from Fort Lee. We had several high school bands and an antique car club and even the local radio stations as participants.

One of the vice presidents made a recommendation to set up a reviewing stand in front of the president's building. When the grand marshall approached the reviewing stand, we stopped the car for him to be seated and watch the parade go by. The chief of police also joined the governor and the president. Needless to say, we had good police protection all around. The parade was a total success. Morale on the campus was sky high.

A member of the faculty, Sam Rhodes, and I worked to get students who had deficiencies in English, math, or any academic deficiencies into the Title III program, a federal program designed to support educational quality and management of financial stability on college campuses. Each year VUU received grant money for the program. We tried to insure all the students were aware of it. From my experience at FAMU, I knew the programs to help students were important.

Graduation rolled around. I was happy because I had watched the students I inherited mature, whether they played sports or they were in the Greek organizations. I was sad. I knew I wouldn't be seeing them anymore, but I knew I would be starting with a new class.

The office of student affairs sponsored two big graduation week events, a senior breakfast and the president's reception. The breakfast allowed me to speak to the parents and seniors to recap their four years and give them a good send-off. At the president's reception, seniors and their parents went through a receiving line to meet Dr. Simmons and the first lady.

When classes ended, we had a two-week break before summer school started. I realized I really missed the students. With Walter Miller, I planned some of the much-needed repairs on my buildings. There was still a lot that needed to be done, although I'd been working with him throughout the year. In summer school, we would not have a large student body, so we congregated women in one or two buildings. We could work then on those vacant buildings to repair all the deficiencies. We consolidated the men on one floor. The university services people, the contractors, could work without interference.

The president said, "Mr. Bryant where are we going to get the money to do all this work?"

"We will get the money," I answered. I'd work more budget line management magic.

When the fall class arrived, my goal was to have a renovated dorm, Huntley Hall, which I wanted to make a co-ed honors dormitory. There had been no co-ed living on the campus when I took over as vice president for student affairs. I outlined my plan: We'd select and invite the students who had strong GPAs. I believed they deserved our trust. The women would be on the bottom two floors and the men on the top two floors.

President Simmons said, "You don't know how to say no, do you? You are constantly thinking of ways to spend money, but in a good way."

"Sir, it's what I've seen in my lifetime. It's what I would like to see if I was a student here. We are raising morale and school spirit at this institution and giving the students their money's worth here at Virginia Union University."

Chapter Sixteen

A Parent's Worst Nightmare

While I was succeeding in academia, so were my children. Young Wil and Lisa were doing all that we asked of them and more educationally. Wil had decided after two years at the University of Southern California he wouldn't stay in electrical engineering. It didn't rub him the right way so he decided that he would give up his four-year ROTC scholarship. I was slightly disappointed, but I always wanted to support him. In Los Angeles, he saw a different world from his military family childhood. He chose to take a different journey and hopefully pursue medicine or business. He researched college costs and considered Howard or Florida A&M but found out as a California resident he could get a great education for a low cost. He told us he decided to change schools and go to California State University at North Ridge to pursue a degree in the health sciences arena. He thought he might want to become a physician. Wil had always worked hard and saved his money, starting in

England where he got a lot of small jobs, mowing the lawn for neighbors and babysitting. Emily established a savings account for him and most of the money he earned he put in the bank. With his discipline we knew he would finish his last two years and get his degree. He graduated in 1988 and he began his search for work.

At nearly the same time, Lisa was excelling at Princeton. She entered Princeton in 1989 and graduated in 1993.

At Princeton, Lisa thought the university needed a cheering squad. So she met with President Shapiro and convinced him to give the squad a budget of $6,000 for uniforms. Lisa captained the squad, just like she had in high school.

Lisa was also a charter member of the Greek organization Delta Sigma Theta, a well-known Black sorority. Emily was not, at that time, a member, so for the first time Lisa couldn't communicate with her mother about her activities because they didn't allow discussion of its internal workings. So anyway she crossed the burning sands and continued her natural leadership among her peers.

When it came time for Lisa to graduate from Princeton, our family, Emily and myself, and our friends were excited. We went to graduation because, as well as earning her degree in sociology, Lisa was being commissioned as a second lieutenant. She had taken full advantage of her ROTC scholarship. Naturally I was very proud she had followed in my footsteps. She graduated with cum laude honors in June 1993. As part of the ceremonies at Princeton, the colleges went to their respective departments and began their awards presentations. The chairman of the department of sociology announced the second highest award for a senior thesis in the department of sociology in 1993 went to Lisa Nicole Bryant. Her senior thesis was entitled *The Army Family*. Of course we were thrilled. I almost jumped out of my seat, and I wondered why Lisa hadn't let us know.

When I asked her about it after the ceremony, she said, "Dad, I wanted to surprise you all. To let you know I'm doing more than cheering and having fun in the sorority."

Shortly after graduation, Lisa was notified she had been selected to be a cadre member at the ROTC Summer Camp at Fort Bragg, North Carolina. I really didn't want her to go to this assignment with so many inexperienced soldiers, and she didn't want to go. As a matter of fact, I had called the camp commander, who I knew, and asked if there was any way if we could get her out of the assignment. The answer was no. Anyway, I left it alone because I knew how the Department of the Army worked. After the summer camp she was scheduled to go to Fort Leonard Wood, Missouri, for the engineer officer basic course because she had been commissioned as an engineer officer.

Seeing our daughter launch into the world was a very happy time for Emily and me. I wanted to take her down to Fort Bragg for ROTC Summer Camp. Lisa allowed me to drive her down in her car so we could leave it with her. She could drive back in the car when summer camp was over.

We pulled into the sign-in station at Fort Bragg, I was getting ready to go in with her, and she said, "No, Dad, you stay right here. I'm going to do this alone. I'm going to be sworn in for active duty as a second lieutenant. I know you know how to do all, but I want to do it myself." I honored her wishes and stayed in the car and let her get her instructions to begin her leadership role in six weeks of summer camp.

Early in the camp, we had a chance to see one of their parades. How proud I was when she was selected as one of the platoon leaders to participate in the parade at camp graduation for the outgoing class. When she was leading her platoon by the official stand, we were sitting in the audience behind the official party. When she came by, it was the responsibility of the platoon leaders to render the hand salute to the reviewing stand. When she did, something told me to stand up and salute and I did. I know she saw me. I'll never forget her beautiful face, her dignified march.

The Fourth of July was approaching, and Emily and I decided to visit her family in Savannah. On our way we stopped to see Lisa and had lunch with her. We came back through on July 5 and stopped again. Lisa took us out to an early dinner before we headed back to Northern Virginia. It was a fun family dinner. Little did we know, that was the last time we would see her alive.

On July 10 in the early morning hours, around 3:30 a.m., our phone rang in our home. The person on the other end, camp commandant Gen. J.J. Johnson asked to speak to me. That's when we learned Lisa had been shot. He said she had lost her life, and they were busy trying to find out who did it. We were devastated. I called my oldest brother Alvin and I called our son. We began the process of getting ready to go to deal with her murder at Fort Bragg. We told another friend of ours in the community, Erv Birmingham, what had happened. He and his wife Sally came to our home because when word got out, we knew that our phone would be ringing off the hook. Erv and Sally Birmingham came over to man our telephones. My oldest brother Alvin called back and said for us to meet him in Richmond, Virginia. We parked our car so he and his step-son drove us to Fayetteville, North Carolina.

When we got there, as you might imagine, we found chaos. All I wanted to do was find who did this and get this man and take him out. I believe I had lost it at that point. I'm a peaceful person, but after Vietnam I knew I had Post Traumatic Stress Disorder, PTSD. This murder of my daughter of course triggered my PTSD. When we arrived there, I couldn't bring myself to identify my daughter's body. My brother supported me; he went in and identified the body.

A terribly hard task happened the next day, Sunday, when we had to pack all of her belongings out of her room. I had to drive her car filled with her things back to our home in Fairfax Station. This was harder than anything, losing my daughter, harder than losing a brother, harder than fighting Viet Cong in Vietnam. We had trained her, given her all the information she needed to handle herself. I just could not stand it. In a state of shock, I pulled into our garage, and I couldn't unload her car. My friend Erv told me to go in the back room and relax as best as possible. He unloaded her car and put all her personal belongings upstairs in her room.

On my own, I was busy trying to find who actually murdered Lisa. I even hired a private detective. I gave him several thousand dollars to find out who committed this heinous crime against our daughter. I wasn't myself. I was always a nonviolent person growing up. I took pride in the fact that I have never had a fight with anybody outside of my brothers in my life. To this day, it remains that way. I raised my children to be the same way.

The military police at Fort Bragg contacted us and said they had a suspect. What we found out a few days later from their investigation was that the man was inebriated and had seen Lisa earlier that evening. We heard maybe he had asked her to dance at one of the local clubs on the post right at Fort Bragg and she had refused him. We left that rumor alone.

This man had sneaked up on my daughter when she was on the phone at the end of the hallway of her building, talking to a friend. She was there because she didn't want to disturb her roommate. He came up on her blind side, with a gun drawn, and forced her to go to his room. At summer camp they had officers and NCOs living in the same building which was a no-no when I was active duty in the United States Army. It is our understanding when she was forced into his room by this murderer with a gun drawn, she determined he was going to rape her. She must have tried to escape.

The military police got the man who murdered our daughter. I received a call from the provost marshall who told me they had the murderer. They believed his DNA matched what was on the weapon and on my daughter. They sent this evidence to the forensics lab to verify their findings.

Immediately after the shooting had happened, the military police put out over 100 military policemen to find the weapon. The man had fled the building right after the shooting. The investigator conveyed to us his attempt to get rid of the weapon on post. He'd thrown the weapon, one of his two weapons, into the trees. They found the gun, and it had been recently fired. Events moved swiftly after that.

Her murder splashed onto national news, primarily because of my friendship with the chairman of the Joint Chiefs of Staff, Gen. Colin Powell. As a matter of fact, it was in most of the newspapers nationwide—*New York Times, Washington Post, L.A. Times, Chicago Tribune.* News reporters swamped my home and yard and I had to push them back. One reporter suggested the reason the Army locked this man up was because of my friendship with Gen. Powell.

I stopped all those reporters right there. "Stop right now. If you ever mention his name again in association with my daughter's murder, you will have no access to me." They honored my request.

Hundreds of our friends came to our rescue to help us through that very difficult period in our life. I was still in shock. I had fought in the war, I had jumped out of airplanes, I'd been on the DMZ in Korea, I'd seen and done just about everything, but nothing like this had ever hurt me so deeply. Losing a child who had done so much in her short life was the worst I'd ever lived through.

The funeral was set for six days later. I don't know how we pulled it together that fast, but our many friends helped.

The evening before the funeral, the night of the wake, Gen. Colin Powell came in, dressed in civilian clothes, to convey his sympathy.

At that time I asked him, "Sir, with all due respect, please do not attend the funeral tomorrow because of what the press might say and what they have said up to this point."

"I understand, Wil," he said. "If you want, Alma will come."

"Alma should be free to attend." Sure enough, his wife Alma was there at the church that day.

On July 16 at the Memorial Chapel, Fort Myers, Virginia, Lisa received full military honors and burial at Arlington National Cemetery. Mind you, she was only a second lieutenant and had only been on active duty for a month. Her funeral was done with the highest honor, in my view the same as they had done for President Kennedy. Her casket on the caisson would be pulled by a single horse.

The day of the funeral in the Memorial Chapel, our family assembled in the ready room. There was standing room only in the beautiful chapel. I have never been inside for a funeral, only for a service. As we were waiting in the ready room, getting ready to take our seats in the church, the chief of staff of the United States Army, Gen. Gordon Sullivan, came to convey his sympathy. Although retired, I wore my uniform for her funeral. He came in to convey the sympathy of the entire United States Army.

The general said, "Colonel Bryant and Mrs. Bryant, if ever there is anything I can do, don't hesitate to call me in the future." I know that is the customary conversation at a time like this. However, a colonel was not going to call the Chief of Staff of United States Army with a problem. I could submit that problem to go up to him through channels.

I couldn't hold back—right then—I said, "General, I really appreciate that on behalf of all my family and your being here today. One thing I would like you to do," I took my hand and pointed my index finger at him. "Sir, when I was on active duty, officers and enlisted did not share living quarters. Sir, it is within your power, please correct this situation so something like this won't happen ever again. Because when you have men and women living together in the same building, that's a problem." He said he'd look into it and take the appropriate action. I said, "Sir, thank you very much."

When the general walked out, the under secretary of the Army, John Shannon, walked up to me as we were getting ready to be seated, my wife, my family, and my mother, in the church. He said, "Colonel, you don't point your finger at the Army Chief of Staff."

I lost it. I used some expletive-laced language and told him "If you don't get out of here in five seconds, I'll knock you out and you won't get out of here alive."

My brothers grabbed this guy and they grabbed me and ushered him out of the ready room. Later I learned he was terminated a few months later. What an insensitive jerk.

We decided that we would not ride in the hearse. We would walk behind her casket all the way to her grave site. Lisa's funeral was beautiful and emotional for us. So we came home and I don't know how I did it—the loss of her was about to destroy me internally—but I returned to VUU the next working day to begin my healing from this tragic event.

Lisa was a wonderful child and wonderful young woman who loved us and her brother Wil. At her funeral, young Erika, our granddaughter, was a baby. She doesn't remember any of this. Lisa did have the opportunity to travel to Los Angeles to see Erika right after she was born, seeing Wil and his wife Gail. Lisa had held Erika. I have always believed Lisa transferred her sparkling personality to Erika. As Erika grew I felt Lisa's ability to speak, both oral and written communication skills, had been transferred to Erika.

The man received a general court martial at Fort Bragg, and was convicted of murder. Emily and I both took the stand to provide a victim impact statement. I stared at this man who would did not look at me. He was flanked by two military policeman armed with weapons.

I asked him, "What could my daughter say, what could she do to cause you to do what you did to her?"

He never looked my way. He never uttered a word in the court, and the convictions stood.

Many friends and my brothers were with us in court, and I needed their support. I think I wished I could get this murderer, but I knew I had to let justice and the military court do its job.

A very close friend of our family, Cecily Hunt, traveled all the way from Savannah to be in the courtroom with us. She is very important to us, and a close friend of the family and loved our children. She gave us comfort in that very trying time.

There were 12 members on the court martial panel; 11 of the 12 members of the jury voted guilty, but one person voted not guilty. Under the Uniform Code of Military Justice with that vote, he could be convicted of murder. He got life in prison, but he would be eligible for parole after 10 years. Word was leaked to us a Black female command sergeant major had voted not guilty. We were told her reason was in her view there were too many Black men on death row and she wanted to spare his life. I don't know this for a fact and no one in an official position confirmed it, but I believe that's exactly what happened.

For months, news reports about her murder continued around the country. News reporters were still hanging around. We were living through an unreal period. So much was written about her murder because of the nationwide coverage of this horrendous event and about our friendship with Gen. Powell. Lisa had ridden her bike past his house as a kid and she called him Mr. Powell. He had nothing to do with it other than being a friend since 1978.

Even after the trial, the publicity continued. We were contacted by an assistant producer from Hollywood who wanted to do a documentary on Lisa's life. They offered us money. I said, "Absolutely not, we don't want to do anything to make any money off our daughter's death. Her record stands on its own."

The Oprah Winfrey show called. One of her assistant producers called, saying Oprah wanted us to be on her show. She felt what happened to us had happened to other families in the United States. Maybe there was something that we could say could help other parents get through their own tragedy. *People* magazine ran a couple of big articles on her death, quoting me, quoting Emily. I believe this was why Oprah's show picked up on it and made the offer to us.

After two or three phone calls from an assistant producer of Oprah's show, we relented and decided to appear on her show. This interview was the only thing we did publicly because we felt maybe it could help others who had similar experiences. They flew us to Chicago, put us up in a hotel, and took us around in a limousine and gave us a tour of the Harpo Studios, I'm sure, all trying to temper our anxiety and hurt.

On the day of the show taping, I was curious how it would be done. Of course I had watched Oprah's show, but I had never believed I would be on a live TV show. Emily and I were the only two people on the set with Oprah. The show was filmed one week and shown the following week. There were two other families on the show but they were in remote locations. They had suffered similar acts of tragedy. The show was professionally done. It was my first time meeting Oprah in person. She is all that they say she is: kind, professional and caring.

Between commercial breaks, Oprah started crying and I told her, "Please don't do that because the commercial will be over in about two minutes and we have to come back on."

It was like a dream—I can't believe we did all this. "5-4-3-2-1, you're live," the cameraman said. That was the kind of attention that Lisa's death brought to us.

We decided we would honor her memory by establishing an endowment at Princeton University. Once it reached the full amount we had established, annually a cash award was given to a student who met the criteria for the scholarship. It soothed us to help a student in Lisa's memory.

Every year we receive a letter from Princeton University to inform us of the recipient of the award. The Department of Sociology also established an award that is in existence to this day, the Lisa N. Bryant award. During Princeton's commencement weekend graduation, we are invited each year to present the Lisa N. Bryant award.

I went back to work Virginia Union to continue as vice president of student affairs to help me get through this, a parent's worst nightmare.

Mr. Secretary—From the Potato Fields to the White House

Chapter Seventeen

Working for Governors

In Richmond, George Allen was running against Mary Sue Terry to succeed L. Douglas Wilder. I watched the race closely because of my membership on the Richmond Republican City Committee. George Allen was committed to being tough on crime and to improve education. In an exciting election, George Allen came from behind and won the governorship by 300,000 votes.

At that time, I still had no ambition for a position in state government because I was happy at VUU. The Republican City Committee invited me to various events, some involving Candidate George Allen. I had not met him personally, but I shook hands with him. I didn't know at the time but later learned he knew of me. He was aware of my career and our family's personal loss.

Two days after the election, I was in the president's dining room when President Dallas Simmons said, "George Allen won. It wouldn't surprise me, Wil, if you don't take a position in his administration."

"Mr. President, I have no desire to do anything like that. I'm very happy here as one of your vice presidents. It is my intention to stay right here."

On the third day I received a call from the governor's chief of staff on his transition team. The governor-elect had asked him to call me to offer me an opportunity to compete for a high-level position in his administration. I was surprised and asked what position was. He answered the secretary of education. They knew all about my resume almost like I had given it to them. He asked me to come down right away for an interview and not to tell anybody. I guess Richmond is a city of secret meetings. I let my executive secretary know I'd be out for an hour or two. I'd have my cell phone if there was an emergency.

Six young people in the room interviewed me. They spent an hour asking questions about education, my background, my schooling in the military and what kind of assignments; they wanted to get a measure of me and a personal assessment of my career. They must have been impressed because the chairman asked me to take a seat in the hall as the governor-elect would like to interview me. This appointment was fun, but I thought for certain I would not be selected. Again I was happy at Virginia Union University because I never thought I'd be a vice president of anything.

Here came this tall guy in cowboy boots. He had a wad of tobacco in his mouth. "Hi, I'm George Allen," he said. "Come on in, Wil."

I thanked him for giving me the opportunity to interview for a position. I said, "First of all, let me tell you how proud I am of you for what you've done and the fact you're the son of George Allen, former coach of the Washington Redskins, one of the over-the-hill gang." He chuckled at that. I mentioned, "I first came to Washington, D.C. in 1970 when the over-the-hill gang was getting started."

We talked football for about 10 minutes and then we began talking about education and what he wanted to do. He wanted to hear my views on education. I summarized my education, both my degrees and my military training, much of which he already knew. I gave him some details how I got where I was, emphasizing the early years, our poverty and my single-parent household. I told him I believed education was central not only to knowledge but to accomplishment. With a good education a person can do anything in this world, my mother had taught us. I said I am who I am today because of the values inculcated in me by my mother and my two older brothers. We talked for about an hour and a half. He thanked me for taking the time to come over. He knew how busy working at a university can be, there was never a dull day and certainly never a dull day in Student Affairs. He'd interviewed over 20 people so far. We shook hands and I said it had been an honor. Back at the university, I forgot all about it.

Two days later, Jake Timmons, the governor's chief, called again. The governor had asked him to call and offer me, not the number one position in the state, but the number two position in the state. I said, "What!" He continued the number two position, the deputy secretary of education. I was honored. The governor had decided on someone else to be secretary, but he wanted me on his team. He would have two deputy secretaries, one for K-12 and one for higher education.

At this point, I told him, "I don't make unilateral decisions. Let me call my wife and discuss this with her and I'll get back to you."

"Colonel Bryant, please do that right away because the train is moving kind of fast."

"How fast?"

"Call me within one hour," Timmons said.

Emily was working for Fairfax County government as a director so it was easy to call her direct line.

"Hey, I've been offered a position as deputy secretary of education in Gov. George Allen's administration. I think I'm going to take it."

"I assume you talked with them about the compensation package," she said.

I told her yes I had, although I hadn't. I still hadn't learned to do that. The salary at Virginia Union wasn't that high, but it was an add-on to my military retirement. Of course, I quickly found out and did the math as always. I called the chief of staff and told him I accept. That's when I asked him what the compensation package was. He told me, and it was a highly satisfactory package. I mentioned money was not the only thing, but I had to look out for my family.

He needed me to come over right away because Gov. George Allen's protocol was that he would personally offer me the position and receive my acceptance. I told him I'd be right there. Again he wanted me to keep all of this secret.

At the transition headquarters. Gov. Allen said, "Wil, I would like to offer you the position of deputy secretary of education to serve in my administration for four years."

"Governor," I said, "I humbly accept this position." We shook hands on it.

"We are having a press conference tomorrow at the headquarters here. We want you and your family, and whoever else you desire to be there."

"I'm going to have to brief the president of Virginia Union University."

"Wil, I want to keep this secret until we announce it."

"Governor, it would be totally inappropriate for me to accept a position, have it announced to the world without letting my boss know. It would be highly disrespectful, but I'll ask him to keep this secret." The governor reluctantly agreed.

I hurried back to campus, going straight to the president's office. I asked his senior executive assistant if Dr. Simmons was in. I needed to see him for a few minutes. She was going to hold me off because I didn't have an appointment and pressed me what was it all about. I told her I didn't need an appointment and I'd discuss the subject with him. Please let him know I was here. She finally said all right and I went in.

Inside, he and I exchanged greetings. "What's going on, Wil?"

"I have good news and bad news." Then he used an expletive that I can't repeat. He asked what was the bad news. "The bad news is that I'm leaving the university to join George Allen's team as deputy secretary of education."

Dr. Simmons displayed a burst of energy I had never seen in all the time I had worked with him. He jumped up and said, "Goddammit, I knew you would take off and take a position in Allen's administration."

"Mr. President, the good news is it will give me an opportunity to see education through a different lens. I'll be right here in Richmond. I'll be available for whoever follows me in this position to help that person learn the job."

"Wil, you've done so much at this university." He repeated all of my accomplishments and asked if I wanted to give that all up to work at the state level. "Do you know, if you do something wrong there, they'll have you out of there within two hours. The state police will have your desk cleared out and escort you off the grounds."

"Mr. President, you can do the exact same thing here. I am not tenured. All it would take is for a female student to walk in your office and say I sexually harassed her, and you would terminate me on the spot."

"Wil, I would never do that to you."

"Stop," I said. I wanted to help him see my side of it. "You came here from another university here in Virginia to become president. You had only been there for three or four years, but you took an opportunity to advance your career. We both graduated from college in 1962. You took the academic route, I took the military route, but I have come back to education. This position is an opportunity for me and my family to do better, bigger things. "In the interim I will be able to help this university. As a private institution, you are entitled to TAG money (tuition assistance grants), and I will be in a position to make sure you get all that you are supposed to get and then some." He agreed that would be true. "Mr. President, I made the decision. I'm going to leave and I want to thank you for giving me the opportunity to be here to be your vice president for three and one half years." Mind you, this was in November 1993. The governor was going to be sworn in January 1994. I said, "President Simmons, one final thing, I'd like you and the first lady to attend the press conference tomorrow when governor-elect Allen announces and other key members of his team. I'd like you to be there."

"I don't know about that," he said.

I said, "Please be there."

I called Emily to contact our son to have him be there for the biggest professional move that has ever happened in our family.

At the same time I received a call from governor-elect Allen and he said, "Wil, there is one other thing, I'd like to ask before

Mr. Secretary—From the Potato Fields to the White House

you come over to be sworn in. Do you have any problems working for a woman?"

"I don't have any problem working for anybody as long as they give me a job to do and allow me to do it, and not micromanage me." Suddenly I realized that in my entire life, especially the military years, I had worked with no women. The governor said that's what he wanted to hear. For the press conference, he needed me to prepare a one-page statement of acceptance to be ready after my announcement as deputy secretary of education. The one-page statement had to be cleared by his press secretary.

Over the years I've never had to clear anything I had written, except when I was in the military. The governor explained in the political world I had to have my comments in writing, and I had to stick to the script. I told the governor-elect that I had invited President Simmons and his wife. I was going to get off script and thank those individuals if they came.

At the press conference, Gov. Allen introduced the secretary of education, Dr. Beverly Sgro, who had been the dean of students at Virginia Tech. He announced her first and gave her an opportunity to give her statement. Gov. Allen next introduced me as the deputy secretary of education. He said a few words about me and offered me the podium. I had never been in a political environment that intense. There must have been 150 cameras and TV and print reporters from all over the state.

I began, "First, thank you, Governor, for putting your faith and confidence in me to appoint me to this key position. I also want to thank Dr. S. Dallas Simmons, president of Virginia Union University and his first lady, for giving me the opportunity to be his vice president of student affairs. He taught me a lot and I'll be eternally grateful to you, Mr. President. Were it not for you, I would not be here today to accept this position."

At that time the entire press stopped their cameras, and they gave me a big round of applause. It was unusual for anybody getting an appointment like mine to get applause from the press.

Never once did I ever think that I'd work at the state level, although I knew I could do the job. Dr. Sgro selected Steve Janosik as the other deputy secretary of education. We got right on the job in early January because the general assembly was already in session.

Once we got started, the secretary made the decision to use Steve Janosik to be responsible for higher education and I would be responsible for K-12 education. As I reflect on the appointment, this was a blessing in disguise. In my view, most of the problems in Virginia's education system were in the K-12 system. We were spending a lot of money, $16 to 17 billion a year on K-12 and higher education. Being deputy secretary, I wanted to learn as much as I could to properly advise the secretary.

Gov. Allen was well organized. He was younger than I, but he had an outstanding support staff. His chief of staff, Jay Timmons, was a perfectionist. I was really surprised after having been on the military side and working at the university to see this level of civilian excellence at the state level.

We were busy attending General Assembly committee meetings. Subcommittee meetings started at 7 a.m. daily. I tried to be there for every one of those meetings. Then there were the full committee meetings. If the governor was backing a bill, he expected his deputies to lobby the members of the House of Delegates and the State Senate.

I had never seen partisanship in action until I began in this position. As a former military guy, I have always believed in trying to do what is right, which was providing a quality education for every child in the Commonwealth. Political partisanship was obvious when I went to the various delegates' offices seeking their support for a bill. In general, Republicans and Democrats received me cordially. Gov. Allen was easy to work for because he worked through his chief of staff and his cabinet secretaries. Bev Sgro and Steve Janosik and I were all on a first-name basis. She and I were the same age and we became great friends. Steve was a couple of years younger, but we all worked together as a team.

The secretary received speaking engagements every week. Many times Secretary Sgro asked me to pinch hit and take a speaking engagement for her. I was glad to do it because it gave me an opportunity to travel around, and people got to know me. I spoke at high school graduations and attended events at the elementary school level. You name it, I did it because I considered it all a learning exercise.

Perhaps the most substantial event was the formation of the Governor's Commission on Education. The standards of learning were insufficient, not really standards at all. His commission would revise the standards to improve the education of our children in the K-12 system. The commission became my primary area of responsibility. We held over 50 town hall meetings all around the Commonwealth of Virginia to get input from parents, students, universities, corporations and the business world, schoolteachers: all facets of society. Over 5,000 people spoke in our hearings. I attended all 50 town hall meetings, regardless of where they were. They could be held as far away as Northern Virginia or Northern Neck or Southwest Virginia. At the end of those session, the commission had the information to create a quality outcome.

In the end we revised standards in the core subject areas of math, science, English, and social studies. Next we needed assessment instrument tests to test the students on those new standards. We contracted with an assessment company to create a new testing instrument. Immediately there were some complaints from some sectors that our teachers would probably be "teaching to the test." We disagreed. They were going to be teaching to the commission standards.

Again the political environment was toxic. In the Army I had learned the value of honesty and integrity and being above reproach in everything I said and did. I played no politics in education because I knew how important it was. I never would have obtained positions in the military and at the university had I not taken my education seriously. When I met with members of the legislature, they knew I took no crap from anybody when it came to our children's education. If some one was incorrect, I would correct them in a kind and tactful way. I'd emphasize what the government was attempting to do, but we were not going to play any politics with education. There was no room for it.

During this period when we worked on the Standards of Learning, I took two significant trips. Our first trip was to Russia. Virginia had partnered with a local school district about 45 miles outside of Moscow. The secretary decided she would take a delegation to visit the school district. We were interested in their teaching methods in the core areas of math, science, English, and social studies, but primarily math and science. Bev chose me to accompany her on this trip along with a team from Virginia's Department of Education. We flew into Moscow.

As a former army officer, I was a little concerned about my status there. Although it was no longer the Soviet Union, but Russia, in my view, it was still the Soviet Union. I had to be certain there was no indication of my past association with the military. The secretary told me before going, she wanted me by her side, so she didn't want to be separated from me. With my military background, she wanted me to watch out for her. We arrived and got through customs, forms and lines and delays and more forms, a harrowing experience, and then our Russian counterparts split us up. The secretary went to live with one team and I was sent to another area, hosted by a principal.

This Russian principal had been in her position for almost 38 years. It was a rude awakening for me to see her accommodations with its poor plumbing and indifferent electricity in ugly concrete apartment blocks. These people worked their hardest to do what was best for their children. It reinforced what I had thought all along. The resources of these countries went to defense and very little was spent on education. We promised them a return trip and they did return that third year. The same lady I stayed with came and visited Emily and me in Virginia.

The second major trip was near the end of the administration, First Lady Susan Allen led a delegation to South Africa. We used those opportunities to review their teaching techniques and their teaching in those four core subjects, but also to drum up business. The secretary of commerce and various key leaders from the governor's staff also came along.

We were there eight to ten days in 1994. Apartheid had ended. We were scheduled to meet with Nelson Mandela. Prior to our arrival, a region of the country had suffered severe flooding, so he was not available to see us. We visited Pretoria and offices there and Johannesburg. We also visited Soweto. It was very tough to see how the people lived there, the poverty and tiny houses poorly built, no public parks or amenities.

Our last stop was Cape Town. Before traveling there, I looked to see who was on the embassy staff. The chief counsel at the time was listed as a Bismark Myrick. I was so surprised that I knew this gentleman. When I was in the Army, there was a Lt. Bysmark Myrick. He and I rode the Pentagon bus home to Burke Centre. Sometimes two to three times a week we would be on the same bus. His son escorted my daughter to the Debutante Ball.

I called the embassy. Sure enough, it was my friend Bysmark. He greeted me fondly. I told him we were in country and headed his way tomorrow. He said he'd put out the red carpet out for us. He couldn't believe I was in South Africa. I told Mrs. Allen I knew the U.S. chief counsel in Cape Town.

"You do?" she said. "When we go, I want you right next to me."

"Mrs. Allen, with all due respect, protocol dictates that the secretary of commerce be with you." I was just a deputy secretary. She said she didn't care anything about protocols. If I knew the guy, I was to be with her.

When we touched down, they brought out the stairs to the plane for us to disembark. Bismark literally had his team put a red carpet out. He shook hands with the first lady and gave me a big hug. I had good sessions, made good contacts, meeting lots of educators. They put us up in a beautiful hotel. Cape Town flowers were so beautiful. The night before we left, the U.S. Consulate sponsored a farewell dinner for us. I got a dose of how politics works in embassies. I really had no personal experience with embassy politics and structure, except on my War College visit to Kenya. I had a good feel for embassy work via Bysmark, a retired Army officer but now in a different role as the top American in Cape Town.

At the dinner Bysmark said, "We are going to have some of our finest wine for the Virginia delegation."

I leaned over and whispered, "Excuse me, Mrs. Allen, the governor has 10 rules for his cabinet officials when they are on official travel. Rule No.2: We should not consume any alcohol. I need to bring that to your attention."

"Wil, I appreciate your insight, but tonight we are going to drink wine. I have already cleared it with the governor and it's okay. You're sharp, but I want you to test the wine." So we did.

We returned to Richmond. It was a very valuable trip for me because I learned about their educational systems in Russia and South Africa and gleaned ideas for use in Virginia.

Chapter 18

Mr. Secretary Goes to the White House

Another governor's campaign was heating up between Atty. Gen. Jim Gilmore (R) and Lt. Gov. Don Beyer (D). A lot of people knew Beyer because he owned a big Volvo car dealership in Northern Virginia, but Jim Gilmore was elected governor by a resounding margin. I knew the attorney general and I wanted to continue serving in the education department because, in my four years of speaking engagements, commissions, and the town halls for the Standards of Learning, I had met many people and I had learned so much. I didn't realize at that time it also increased my visibility in the Commonwealth.

When Gov. Allen was getting ready to leave office, we were advised if we wanted to continue to serve, we should write a letter to the governor-elect to let him know which position we wished to be considered for. I loved my boss, Bev Sgro. She never micromanaged and instead gave me opportunities. Under her leadership, I discovered I could work well with a member of the opposite sex as my boss. I submitted three letters to governor-elect Gilmore, requesting one of two positions outside education and one to retain my own position as deputy.

By early December I had heard nothing from the governor-elect. To take a break from politics, Emily and I took a trip to Atlanta. In his transition, Gov. Allen had already selected his team by this time, but Gov. Gilmore took more time making his final selections. As we were driving back from Atlanta, I asked Emily to call home and see if there were any messages on our voicemail.

We put it on speaker in the car heard a message from the transition team, "The governor-elect would like to interview you, regarding a position in his administration." I spent the weekend wondering and hoping.

Monday morning I went to the meeting at 9 o'clock. There were only two people in room, the governor-elect and his chief of staff.

The governor-elect asked, "Wil, first of all I noticed that you never submitted a letter asking to be secretary of education. Why not?"

"Governor, my boss, Secretary Sgro, in my view, has done an outstanding job. I didn't want to compete with her for the secretary position. That's why my requests were for other positions. I figured you would be reappointing her."

"My read on you is just as I figured. You are very loyal," he said. "I've decided not to reappoint her. I want you to be my secretary of education." I was glad I was sitting down because I might have passed out. He continued, "I have observed you over the previous years. You've spoken at universities and K-12 schools. You've attended board meetings. You've done everything. I think you will be a fine secretary of education. I want you and your skills on my team. I think you can handle yourself well before the General Assembly and present my education policy initiatives in a highly professional manner."

I thanked him. He asked if I had any questions, I had two comments to make. "Sir, first of all I've done some research on salaries for Virginia's teachers. They are being paid below the national average. If I'm your secretary of education, I would like to see us bring the salaries of Virginia's teachers up at least to the national average or above."

"That's a tall order," the governor said. "We will work together, and we will try to do that." I knew he planned to send the money from Virginia state lottery directly into K-12 education.

"Governor, no doubt you know, the legislature is supposedly directing the lottery proceeds to education, but they are not. Those funds go into the general fund and can be used for health-care services, transportation, or public safety."

"You're exactly right," he said. "We're going to change that. I recall the first lottery check in excess of $300 million went to K-12. A formula was used to distribute those funds, a composite index of each local school division's ability to pay. The poorer school division would get a larger percentage and the wealthiest school divisions like Fairfax County or Loudon County a smaller percentage."

"Good," I said, "My second comment. As you know, we have two public HBCUs. Those schools have been underfunded since their inception."

"Wait, Wil," he said. "I know where you're going with that. As my secretary of education, you will be submitting a proposed budget to me every year in September or October. You make your recommendations what these schools should receive, using information from the State Council of Higher Education. You come to me with that."

I wrapped up, saying, "That's all my concerns." He told me the swearing in ceremony was the next day and to invite my family here.

Emily held the Bible for my swearing in. At the ceremony, one of the first persons to congratulate me was my oldest brother. He said, "Mr. Secretary, congratulations." He shook my hand. I really couldn't believe I was the secretary of education of the Commonwealth of Virginia.

What a great day, all 10 cabinet secretaries were sworn in. The next day, we immediately went to work because the General Assembly was already in session. Much was expected of me and my department. I had the largest secretariat in the state government with a bi-annual budget of $17 billion. I worked regularly with the State Council of High Education, which was primarily responsible for providing policy and budgetary guidance to the governor and the House of Delegates and the State Senate. All that information made its way through my office, which I had to digest, filter, and weave into my initial report each year to the governor. All the different educational institutions, 37 educational entities, were in my secretariat, a tremendous amount of resources provided by the people of the Commonwealth and a tremendous amount of responsibility.

The first order of business was to bring in the two deputies who had been selected to give them guidance. Dr. Cheri Yecke had accomplished many things at the State Board of Education. She was my deputy for K-12. Robin Zink was my deputy for higher education. I selected David Adams to be my special assistant. He wrote many of my speeches. He would also drive me during our travels in the Commonwealth. Each cabinet secretary was issued a car for their use. The governor also had three corporate jets so if we had to travel two or more hours outside Richmond, we were free to use one of those jets whenever we needed to. We couldn't waste time on travel. The cabinet secretaries were going to be very busy. Our work started out in a hurry.

That Monday morning after I was sworn in, I had to take care of a personal issue with a member of the house of delegates who had made a negative comment to the press that I appeared to be minimally qualified for the job. I took issue with that. My special assistant David and I went over to the House of Delegates to visit this person.

"You made a comment to the *Washington Post* that I appeared to be minimally qualified," I said.

He told me to "Hold up, hold up."

"Nope, you hold up. I already talked to the reporter and checked his tape recorder and he confirmed you actually said that. Delegate, let me just say this. This is my first day on the job. I'm a retired Army colonel. I fought in Vietnam, I commanded troops in combat, I've jumped out of airplanes, and I have traveled all around the world. You are aware of the degrees I have earned. I have more qualifications than any 10 or 15 members of this House of Delegates. So please don't make any more negative comments about me unless you have facts to back up your statements." I was fuming.

He said he was very sorry and he looked forward to working with me. I turned and left his office. The exchange was tense, but I wanted to get it taken care of. Of course word spread far and wide—don't mess with Bryant.

In my office I plotted our work schedule. I asked my two deputies to keep me informed. We met once a week formally or when required. I didn't believe in a lot of meetings. If they had something to report, I asked them to convey it to my special assistant. We'd find a time to accommodate their busy schedules. As a deputy secretary, I knew what they were encountering on a daily basis.

Our first order of business was to complete the education budget. The only changes we could make to Gov. Allen's budget were to prepare budget amendments to be added to what was called the caboose bill.

In my first months, I traveled to Virginia State University, an HBCU, where I was invited to give the convocation address. It was held in Constitution Hall, which was in serious disrepair. I made a mental list of the problems.

The governor had a cabinet meeting every Monday morning at 10 o'clock. If he was not in town, his chief of staff conducted the meeting. Right after that trip, I was the first of the cabinet secretaries to give a report. At the end of my talk, I told the governor, I needed to see him privately for five minutes. When the meeting ended, I told him about Virginia State University's problems. He told me to fix the problem via the caboose amendment.

I contacted the area's delegate and the state senator to ask them for funds in the budget caboose bill for Virginia State University to renovate that building. We also included funds for Norfolk State, another HBCU, because we couldn't give Virginia State money and not give money to Norfolk State, which was in the next town. We invited both college presidents to a press conference on the steps of Constitution Hall. I introduced Gov. Gilmore and he announced the amendment to give them monies. Those two HBCUs knew I was serious.

Within the first month of the administration, I invited the state NAACP director and his key staff to meet with me. I wanted them to know who I was, my accomplishments, and also my perspective and my philosophy on education. I asked the state superintendent of public instruction, Paul Stapleton, and three members of his staff to join us. Over two hours, I first did the talking and then gave them an opportunity to ask me questions. At the end of that meeting they understood my plans to help all the children in the Commonwealth of Virginia, not White and not Black, but all the children. It was a good start and I'm happy to say that I had a great relationship with the NAACP. They invited me to speak at their convention and we did events together all around the Commonwealth.

The Standards of Learning had been implemented, and we were moving rapidly ahead. We tested our students at grades three, five, eight and in high school. The first time we administered that exam, the scores were low, but we saw improvement across the board with each year of that administration. Those results delighted me, proving the commission had done outstanding work in its recommendations.

After I was sworn in, I wanted to visit the 16 public colleges and universities in Virginia as quickly as I could. With our huge budget, the largest budget in the state government, I had to have a good grasp how this taxpayer money was spent to make sure we didn't waste the monies.

My driver and I took off to Radford University in southwest Virginia to meet the president. At each university my format was to meet with the president, the chairman of the board, the president of the faculty senate and in some cases, meet with the president of the Student Government Association. I wanted a tour of the campus because these presidents would be submitting budgets to the governor through me and my office. I wanted to see and understand what they were going to be asking for.

Next we drove down U.S. 81 to Virginia Tech, and we visited all the schools in the southwest of the state. At the University of Virginia, I spoke to President John Casteen and his team. We continued until we hit all 16 public colleges and universities. We had nine comprehensive universities in the Commonwealth of Virginia, six doctoral institutions and one two-year program, Richard Bland College, which came under William & Mary.

I'd brief the governor in September or October every year. Virginia governors serve only one four-year term. For my major presentation I included my two deputies and my special assistant. Of course, the governor assembled his entire team, which filled his conference room. I was the person on the docket to give the education budget for the entire Commonwealth. With all the information I had gathered, I could make a decent recommendation to the governor what college A or College B should be given for the upcoming school year. I felt it was a very big job, but I knew I could handle that.

I continued my speaking around the state because it was important to meet with those colleges and universities. The legislature was debating faculty tenure because many felt the system should be changed. If the faculty were not performing up to standard, there should be a post-tenure review set up to assess these professors and take whatever action was appropriate. We didn't make a lot of progress on that issue during our time, but we did make some.

A different example was Christopher Newport University, a transient school at the time. A former United States senator, President Paul Tribble wanted to make it a full-fledged comprehensive university. On several visits, he drove me all over the campus in a golf cart, mentioning all the buildings he wanted to build in order to add more classrooms. My department accomplished those improvements at Christopher Newport.

Gov. Gilmore first chose to bring in Paul Stapleton as state superintendent and he stayed for most of the administration. When he left, I made a recommendation to the governor to bring in Dr. JoLynne DeMary. She and I had worked closely together when I was the deputy secretary. He selected her to be the state superintendent for public instruction and she did another outstanding job.

My position gave me opportunities to interact with other chief state school officers, secretaries of education and state superintendents through the Educational Leaders Council, headed up by Arizona State Superintendent of Educations Lisa Graham-Keegen.

When I was secretary of education, Texas Gov. George Bush was campaigning to be president, and he came to Richmond with his wife. They visited a local school where Mrs. Laura Bush read to a group of school children. Gov. Gilmore had to be out of town, so he asked me to attend that meeting with Gov. Bush.

Bush was a jovial fellow. He was younger than me, but he was a governor, so I gave him respect. He had been briefed on my background. When it was all over, we had a small press conference so the press knew about his platform and then he continued his campaigning.

Gov. Gilmore held five 50th birthday parties as fundraisers for Gov. Bush. The final birthday party was held in Northern Virginia as a big fundraiser for the big contributors. As the conversation would be about education, the governor asked Emily and me to be there. Tickets were up to $25,000 per person to attend, but we received complimentary tickets. I appreciated that.

When we arrived at the party, Emily carried a gift bag. I asked her what it was. Emily retorted, "If Governor Gilmore is having a 50th birthday party, I'm going to bring him a birthday gift."

"Emily, these are not birthday parties. That's just a name or label they gave them. They are fundraisers for Governor Bush."

"No," Emily said. "That doesn't matter. I'm going to give him a gift."

A receiving line formed. Since I was secretary of education and one of the highest ranking in the cabinet, we were asked to go through the line first.

Gov. Gilmore said, "Governor Bush, this is Wil Bryant, my secretary of education and his lovely wife, Emily." We shook hands.

"Governor Gilmore," Emily said. "I have a gift for you."

"I told Emily it was not a birthday party, but she insisted," I said.

Gilmore said, "Wonderful." He looked in the bag and there was a book and a crystal elephant. The governor had a big chuckle over it.

Gov. Bush said something to Emily, and she said, "Oh you sound just like your mother."

"Wait a minute, when did you meet my mother?" he said.

"When my husband was in the Army, he was a student at the National War College. I was invited to the Naval Observatory to have tea with your mother."

We were holding up this long line. I said, "Governor Gilmore, we have to move along and let these other people come through." It was a fun night, a success. I didn't dream I'd be meeting him again.

Also under my secretariat's responsibility, I had a wide assortment of institutions, museums, specialized schools, and the State Library of Virginia. I enjoyed working with five state museums because the museums complemented what was being taught at our schools at all levels in Virginia. The list included Gunston Hall, Jamestown and Yorktown in the Hampton Roads area, and the

Frontier Cultural Museum in Staunton. A very important museum, The Science Museum of Virginia, was headed by Dr. Walter R.T. Witschey, a Princeton University graduate. We worked well together in part because Witschey knew my daughter had been a student there. The fifth museum was the Virginia Museum of Fine Arts. The State Library of Virginia was a big part of the educational process because not only did it serve all the colleges and community colleges but all people of Virginia. I felt it was an honor and a pleasure to work with them.

Also the community college system, schools for the deaf and for the blind, and two teaching hospitals, the Medical College of Virginia at VCU and the University of Virginia Medical Center, all fell under my responsibility. I traveled to the majority of those specialized institutions to try to make sure that we provided them with whatever support they needed.

One of the visits that really touched me was to the two schools for the deaf and blind. First of all, they told me they'd never had a secretary of education visit. What I found there was appalling. The equipment and the school buildings for the children were totally unacceptable, starting with no air conditioning in the rundown buildings. I couldn't wait to let the governor know what I had found and what we needed to do to correct these problems. We needed more Braille equipment and updated reading material. We needed to make the buildings comfortable for these children who lived there on the property. Some children commuted daily on a school bus. I wanted us to upgrade those properties. We put money in the budget to get that done. It was a top priority.

The governor said, "Wil, you are going to take all the money out of the bank!"

I said, "We've got to get this right because the children need it and they can't fight for themselves, so I'm going to fight for them."

My great challenge was presenting Gov. Gilmore's first budget to the House of Delegates and the Senate Appropriations Committee. I had to follow a strict format when appearing in front of the 22 or 23 members of the committee. The scene reminded me of Supreme Court justices in their high backed chairs on an elevated dais. After I presented the governor's budget, I had to take questions from the house and senate members.

Earl Dickinson, the chairman of the appropriations committee who was a Democrat from Louisa, Virginia, introduced me as the secretary of education, the Honorable Wilbert Bryant. I thanked them for all for giving me the opportunity to present the governor's budget.

I deviated slightly from my script and said, "To stand here before you and look at the people in this room, an awesome gathering of legislators—had I not had a military background, fighting Vietnam, jumping out of airplanes and serving along the DMZ in Korea, I might be a little intimidated. But I'm not. When I complete my presentation, I will answer any questions you may have."

I read the prepared budget text and when I finished, I took their questions. I was before them for about four hours. I thought I had done a pretty good job because I had carefully prepared, by taking home two briefcases to study every night. My daily routine was when I came home around 6:30 or 7 p.m. at night, after eating somewhere, I would study like I always did. I would read my material for the next day whether the General Assembly was in session or I appeared before the House Appropriations Committee. I wanted all the information in my head. I wanted to memorize as much as possible. When I received the governor's budget, I knew there was no way I would be able to memorize all of it, but I familiarized myself with the key points so I wouldn't have to look down or turn the page or try to find information, but I would maintain eye contact with whoever asked the question.

When I finished my presentation, one of the governor's assistants was in the room. The print media, the electronic media, and lots of cameras were in the committee room. I didn't know it was being televised in the governor's office and they had followed my performance. The governor's assistant told me the governor would like to speak to me.

Gov. Gilmore said, "Wil, I just wanted to let you know you did a damn good job. Keep it up. You hit a grand slam home run with your presentation."

"Oh I thought you were getting ready to work me over for making some comments that I shouldn't have made."

He told me, "Oh no, You've done a great job. I am very pleased."

Another major organization in the education area who reported to me was the State Council of Higher Education. Higher education required a lot of my attention and a lot of taxpayer money. We had one of the finest universities in America and in the world, the University of Virginia, which was our flagship institution, followed closely by William & Mary, Virginia Tech, George Mason University, Virginia Commonwealth University and Old Dominion University, all doctoral institutions. George Mason had been pretty much a transit institution when I was the deputy secretary of education. Gov. Gilmore took George Mason to the next level. We gave them several hundred million dollars to develop programs, buildings and faculty, so that their per capita disbursement for students was commensurate with the other doctoral institutions. GMU's President Alan Merten lobbied the general assembly, the governor and me for support. I had been acquainted with him when I was deputy secretary through those 50 town hall meetings. Allan Merten was one of first people to appear before the education team to give his views on improving Virginia's Standards of Learning. We became good friends. George Mason is now one of the outstanding schools in the Commonwealth.

In Washington, D.C., early in his first term, President Bush announced an initiative for school curriculum for K-12. As secretary of education, I was sent by Gov. Gilmore to President Bush's presentation at the White House. I, a child of a disabled domestic worker, entered the White House's Roosevelt Room.

President George W. Bush, framed by American flags and under the gaze of presidential portraits of Teddy and Franklin, was to announce his education initiative, No Child Left Behind, to the assembled 50 chief state school officers.

I was determined to defend Virginia's own Standards of Learning, which my Virginia secretariat and governors had carefully created after 50 town hall meetings, listening to over 5,000 Virginians. We had achieved three years of steady improvement in school test scores since our standards were implemented. When the president finished his presentation, he asked if there were any questions or comments.

I immediately raised my hand and President Bush said, "I can't remember your name but I know your face."

"Wilbert Bryant, secretary of education for the Commonwealth of Virginia."

The president welcomed me with a smile and invited me to continue. "While we support your initiative, we will keep our Virginia Standards of Learning because we are making tremendous improvement in our children's education, children of all races."

At the end of the meeting, his staff announced there would be a press conference in the Rose Garden. U.S. Secretary of Education

Rod Paige invited me to join him, two governors and one other chief school officer. He asked me if I wished to speak to the press.

"Yes, absolutely," I said. "My governor would expect me to."

The Rose Garden, certainly the most famous venue in the world for a press conference, was filled with reporters and camera crews. When I was called to the presidential podium, I spoke about our Standards of Learning and how No Child Left Behind would be a supplement to them. After a career of public speaking, I wasn't afraid or nervous. Inside I was smiling, thinking of my Army mentors, my family, my five brothers, my disabled mother, marveling how far I had come from picking potatoes in the fields of South Florida at age eight.

Another highlight of my time as secretary: I gave many commencement and convocation addresses during my tenure. One of the highlights of my time as secretary was when the Virginia State University president called and said they would honor me with an honorary doctorate. I thought that was fantastic. He asked if there was anybody I would like to speak at this convocation where I received this honorary degree. I wanted the president of my undergraduate alma mater, Dr. Fred Humphreys, to be the keynote speaker. The convocation took place on Sept 10, 2001. Of course everybody in the whole world knows what happened the next day. That was my first honorary doctorate.

Chapter Nineteen

Strengthening HBCUs and the Peace Corps

When I arrived as deputy secretary and more importantly as secretary of education, Dr. Arnold Oliver was the chancellor of Virginia's 23 community colleges. It was my goal to gain sufficient knowledge of the schools to make an intelligent budget recommendation for them. I knew Dr. Oliver through the State Board of Community Colleges.

In 2000 to 2001, in the Report of the Secretary of the Commonwealth, Anne Petera wrote, "Over the past year, Virginians have accomplished much of which we can be proud. More than 400 or more public schools have already achieved the status of fully accredited because the students have passed the Standards of Learning Tests. This achievement is remarkable considering the schools in Virginia have until the year 2007 to do so. The students and teachers who accomplished this task are to be commended." I was so pleased and proud.

Our commitment to maintain the strength of institutions of higher learning and reduce tuition and fees enabled more Virginians to complete their education and compete successfully in the job market after graduation.

I did not achieve that success alone. We had a superb staff. June Hines was my executive assistant and Sanita R. Dozier was my administrative staff assistant.

At the same time we were achieving these accomplishments in Virginia, on the national level President Bush's administration was off to a great start. I recall after 9/11, I asked Gov. Gilmore to allow me to compete for a high-level position in President Bush's administration. Gov. Gilmore wanted me to stay in my position while we worked on the budget. We would be working on our last budget before his administration finished. He had spoken with the president and was assured I'd have a high-level position. Gov. Gilmore had blessed me in my promotion to secretary of education. I'm eternally grateful to him.

At one point I traveled to Tallahassee to my undergraduate alma mater to assist in a presentation by U.S. Secretary of Health and Human Services Claude Allen. Afterwards, in the hotel, my phone rang. I wondered who would be calling me. Emily wouldn't be calling unless it was an emergency. It was my executive assistant, June Hines telling me Gov. Gilmore wanted to speak to me right away. She patched me into the governor. He asked what was I doing in Tallahassee. I reminded him we had discussed this trip and we exchanged a few glib comments. He then told me he had spoken to President Bush. He had identified a position for me in his administration, deputy assistant secretary for higher education programs, with a reporting date of December 5, 2001. Our budget would be wrapped up in September to October. What good news! Leaving my team would be sad. I'm sure all secretaries think that their team is the best, but I thought mine was one of the finest.

I began my round of goodbyes because the year was rapidly coming to an end. First I visited the Virginia Department of Education to say my goodbyes to state superintendent of public instruction Dr. JoLynne DeMary, and to the education secretariat; they had worked so hard and successfully for me.

In July 2001 Florida's Gov. Jeb Bush had reorganized the governing structure of higher education. He had chosen me to be a Charter Board of Trustees member for Florida A&M University. Florida was aligning more in line with other states by creating a Board of Trustees for each institution. In Virginia, they were called a Board of Visitors. To be a charter member of the Board of Trustees of my undergraduate alma mater—what a high honor for me. Unfortunately joining President Bush's administration would present a conflict of interest, so I had to submit a letter of resignation to the Florida A&M Board of Trustees.

On December 5, 2001, I traveled to Washington and was sworn in as the deputy assistant secretary of education for higher education programs. I quickly assembled my personal staff. The first person I hired was my senior executive assistant, who was to be my gatekeeper, the ultimate authority on any visitors. I selected Paula Hill and had her transferred from the Defense Department. I knew her from my last assignment in the Pentagon in the office of the secretary of defense. She knew how to handle confidential papers and had great interpersonal skills. I gave her a major promotion when she assumed this important position.

My chief of staff, Alan Schiff, was already in place and he knew the operation of my area inside and out. What I wanted from him was a solid briefing on the five divisions under my supervision. I had heard all kinds of rumors about federal career workers, but I have worn the uniform of my country dealing with soldiers and I believed I could handle the situation without any problem.

My people knew I was a presidential appointee and I wouldn't be there very long. That didn't matter. I was going to do the very best job I could. I represented the taxpayers of the United States. We did the job in accordance with the regulations written by Congress to serve as a guide. Our bible was the Higher Education Act of 1965 as Amended, a book about two inches thick. I read it at least 10 to 12 times to soak in the gist of our responsibilities. From my office alone, we would disperse a little over $2 billion a year to colleges and universities around the country, all kinds of programs directed to helping students. I was eager to begin.

I circulated through all my divisions to introduce myself. First, I met with the division chiefs. I told them I would meet informally with some of their people, not to make them feel threatened or gather intelligence on how well they do their job, but for me to get a feel for the papers coming from the lowest level in that organization to my desk to review and approve. Although I always checked, I had confidence in my division chiefs and in my chief of staff. When I sent any paperwork upstairs to the assistant secretary of education, I wanted to make sure it was correct.

Two schools in the Higher Education Act of 1965 as Amended receive tax dollars from the United States taxpayers. One was Howard University, which was getting $240 million; the other was Gallaudet University, the school for the deaf. It touched my heart because my mother was disabled. I wanted to make sure they had all the supplies, equipment, personnel staff, and support they needed. I assigned another staff member the primary task of taking their budget and scrubbing it to help those young people who were in such high need.

Another task thrown at me, right after I came on board, was reviewing of the Howard University budget. When I worked at Howard for four years, I had seen waste, fraud and abuse. Their budget was submitted to my office and then forward to the assistant secretary for approval and then to the Office of Management and Budget in the president's office. Their first budget submitted, in my view, was a disaster. I took one of my GS-13s and I promoted her to GS-14. Howard's budget became her primary area of responsibility. When their budget was submitted to my office, she'd review it, outlining everything that was wrong. I asked her to communicate to the chief of staff what was needed, get it revised by Howard University, and then returned to me. I felt especially good about the changes we achieved.

I was one of two deputies working under the assistant secretary of education, Sally Stroup. I dealt with the money issues, and the other deputy dealt with the policy issues. The other person I worked daily with was the assistant secretary's executive officer, an incredible individual, Mr. Humphrey Barnes. We had some heated moments because in the federal government environment there was a tendency for people in the civil service to think they knew everything and tell you what to do. I told him I appreciated his guidance, but we would do it my way. He soon found out he was dealing with a stubborn retired Army colonel who knew what he wanted to do.

It was a delight working under the new leadership of Secretary Rod Paige, but I encountered civil service union issues. I had never dealt with union issues before. For any promotion, the union became involved. My department also had quite a few people who were dissatisfied with their status in the U.S. Department of Education because they had been a GS-12 or G-13 for 14 or 15 years with little chance of being promoted. At the same time, they were doing work of those at higher grade level, GS-14 or GS-15. I gave the human resources specialist on my staff a special assignment to do an in-depth audit of the number of GS-12s. Of course, she worried the union was going to get involved and maybe bring suit. I didn't care about the union. I didn't work for the union. I told my staffer to get it done. I met with the union and kindly told them to stay out of our business. They never bothered us.

We had several personnel issues, promotion, jealousy of promotion levels and similar matters. I dealt with them on a problem-by-problem basis. I wanted staff to know I was going to be fair. I still used the old saying, "Trust but Verify." When I made a decision, I brought the supervisor and the person submitting the complaint in to insure they understood my solution. In almost every situation, the person left happy. I gleaned a lot of information from my initial days there, walking around and talking to people and letting them get to know me, just the same way I always did in a new position.

Even though there were over 150 people in my area, I wanted them comfortable in the knowledge that I looked forward to working with them.

In my area, I had the HBCU Capital Finance Program with over $300 million dollars from Congress. Its purpose was to enable public and private colleges to submit requests to get a low-interest loan to build academic buildings. A typical loan would be 10 to $20 million at 1 to 1½ percent interest rate. The institution would get 30 to 40 years to pay it back. Many of the HBCUs weren't aware of the HBCUs Capital Financing Program. Part of my job was to get information about this program out to HBCUs.

One day Sen. Lindsey Graham's chief of staff called me to ask me to travel to Columbia, South Carolina to speak to all their state's HBCU presidents, public and private, about the HBCU Capital Finance program. Senate Graham was out of the country; otherwise he would have been there. I spent one full day with all but one of the South Carolina HBCUs presidents.

At 6:30, the morning after my South Carolina trip, Sen. Graham, who had returned from Africa, called to personally thank me for speaking to the HBCUs presidents. I thought that was noble of him to find time to call a deputy assistant secretary.

The HBCU Capital Finance program posed many challenges for these universities. Sometimes a college or university would narrowly miss out on a grant by losing one and a half points on the grant requirements. Many times, a university president would call their member of Congress to complain about the Department of Education not awarding them a grant, and they'd like to have it reexamined. Unfortunately, there was no appeal process for this program. The first time it happened I went to meet with the congressman. I brought a staff member with me for highly technical questions; I would call on him or her to answer, but I would do the talking. I knew more about the program than the member of Congress and the university. Because I knew the process, I knew by heart the criteria for those grants. I'd instruct the congressman how we would respond to the university. When the college or university called on an interactive video, I wanted the congressman to welcome them, introduce me and my deputy and turn it over to me. I'd explain why the grant wasn't awarded. Next I always told the college or university president that I would send a team from my office to work with them how to apply for this grant, so that during the next round of grants they'd be on the receiving end. That approach worked very well.

At a Friday meeting in St. Louis, when I was participating in a higher education panel discussion, my special assistant came up to the stage and passed me a note that said Secretary of Education Rod Paige wanted to see me on Monday morning at 9 a.m. My experience in seeing something like that it was going to be one of two reasons; either you were going to get fired or something good was about to happen. So I had all night and Saturday and Sunday to worry why the secretary of education wanted to see Wil Bryant. At that time, I had not met Secretary Paige except at the Rose Garden press conference the year before.

Monday at 9 a.m. I went to Secretary Paige, who greeted me with a big smile. I asked what was the nature of this meeting. I even told him he had me worried over the weekend. He said I had nothing to worry about. He wanted me to wear a second hat—take on a second job with a second staff. He wanted me to be his counsel to the secretary for the White House Initiative with Historically Black Colleges and Universities. HBCUs.

"Wait a minute, sir," I said. "Former Ambassador Leonard Spearman is the incumbent right now. What about him?"

"You are going to replace him. As you may know, Ambassador Spearman had an unfortunate situation, a stroke that has slowed him down. His ability to travel around the states to visit HBCUs has been impacted."

"Mr. Secretary," I said, "with all due respect, I know Ambassador Spearman very well. He is a close friend of my family. He mentored my two older brothers at Florida A&M University." I was hesitant about this. Spearman had done an outstanding job as an ambassador.

Paige said he understood my concerns but he'd made his decision. He'd discussed this move with President Bush. They both agreed I was the person to operate those two staffs efficiently.

I told him, "Mr. Secretary, this is going to be tough."

He said he'd take care of it with Ambassador Spearman. I asked when this would take place and he said within a week. I agreed on two conditions: move the HBCU staff from their present location to my space in the big building on K Street. If my span of control was going to be increased, I wanted them close. My second request was to change my personal status as a presidential appointee from SES-4 to have the highest level, SES-6, Senior Executive Service Six. He told me it was a done deal, that he'd see me in two weeks to be sworn in. We stood up and shook hands.

Secretary Paige said, "Don't go out that door, Ambassador Spearman is waiting out there. Go out this door instead. Don't say anything to anybody until I have publicly made the announcement of this appointment." Politics is strange, but that's how he handled it.

I immediately made Dr. Len Dawson my deputy. Dr. Dawson had been president at Vorhees College for 16 years. He had the background and the expertise to serve as my deputy. With those two staffs in close proximity, I'd meet with my assistant secretary of education staff one hour and the next hour I'd meet with the HBCUs staff. I maintained an open-door policy. They could call me 24/7, regardless of where I was or what I was doing. I had four telephones to deal with that.

Later I went to talk to Ambassador Spearman. He had been a mentor to me as well. I must confess I was in tears. I said, "Sir, I had no knowledge of the decision to replace you."

"Wil, I know you didn't have anything to do with this. My physical condition is what caused this. You will do an outstanding job. It is very rare for a person to have two jobs in the U.S. Department of Education with two large staffs. I know you and your family and you can do this." I thanked him for his vote of confidence.

Before Dr. Spearman left the department, at the National HBCU conference, I presented him with a certificate of appreciation from not only the higher education community and the HBCUs community but from the people of America for his many contributions to education. His wife and family were present. We must have had 700 to 800 people in the conference room. I know he was very appreciative.

During my time working with the HBCUs, I worked with Dr. Lou Sullivan, who had served as the secretary of Health and Human Services in President George H. W. Bush's Administration. He was the founder of the school of medicine at Morehouse College. He also had been appointed by President George W. Bush to be the chairman of the President's Advisory Board on HBCUs. The board's duties and responsibilities required an annual report on their status of HBCUs with their recommendations to the president and his team to improve those HBCUs.

I recall joining the advisory board to meet with President Bush. We took a nice photograph and we chatted for about 15 minutes. The president's time, of course, was tightly scheduled.

As we continued working, I visited many HBCUs. At Claflin College, South Carolina State University, Lincoln University in Missouri, Gannon University in Pennsylvania and the University of Maryland Eastern Shore, I gave commencement addresses. I traveled to Western Michigan University at the request of Congressman Fred Upton.

As a high honor, I was also installed as a member of the External Advisory Committee at Princeton University's Department of Sociology. On my visits, I interacted with some of Lisa's professors. It wasn't a policy position that would have posed a conflict of interest. Still, I decided I'd submit my resignation to that committee because I didn't want any appearance of a conflict of interest.

For me a crowning achievement was being the recipient of additional honorary doctorate degrees. At Shaw University in the Raleigh, North Carolina area, I gave the convocation address after receiving an honorary doctorate. I received an honorary doctorate from St. Paul College, Lawrenceville, Virginia, and also from Lincoln University in Jefferson City, Missouri, and another from the University of Maryland Eastern Shore.

My final recognition came from Gannon University in Erie, Pennsylvania where Dr. Anthony Garibaldi was the president. One of my classmates from the National War College was his chief of staff. I had met Dr. Garibaldi before at another education conference. He invited me to come out and give the fall commencement address. I'd never been to Erie. As I gave my convocation address, I was slipped a note, telling me I would be not able to stay, post speech, because the lake effect was about to happen. There was only one plane leaving Erie and it was waiting for me on the tarmac. I was told the propellers were turning and I had to get out or I'd be stuck for three or four days. I received my honorary degree, and told them that I had to leave.

The fog was rapidly approaching. A police escort took me to the airport. When I got on the plane I was greeted with loud applause. The passengers had been told the plane was waiting for some government official who had to return to Washington. The folks were afraid they wouldn't get out of there before the lake effect stranded us all. We lifted off and got above the fog. I wasn't worried. I'd done a lot of flying and I knew if the pilot said it was good to go, then it was. We got to Detroit, but we were snowed in there. So that was the last time I went that direction in the winter months.

A special event, the National HBCU conference, was held each year. All HBCUs in America could participate in the conference, hosted by Secretary Rod Paige in Washington, D.C. We had tremendous participation in all our conferences. We had requests from some universities that in addition to university presidents and key members of their staff, they also wanted their trustees to attend.

One of my HBCU highlights was inviting Colin Powell to speak. By that time he was the United States Secretary of State. I had told the conferees on day one there was a high probability we might get Gen. Powell to speak to the group. While he was not an HBCU graduate member, he knew about our efforts. Plus, I considered him a personal friend and I was going to take advantage of our friendship once more. In the planning stages, I received a letter from him, saying that his schedule would not allow him to be there, but if there was a change in his schedule, he would speak to the group.

On the second day of our week-long conference, I received a call about 7 p.m. from Gen. Powell's special assistant. Secretary Powell had a two-hour gap in his calendar tomorrow morning, if we could work him in. It was a done deal, so I announced Secretary of State Powell was going to come. We had some 700 to 800 people, a full house, in the Marriot Hotel in Crystal City. I had worked with his security detail in advance and was told they were coming in one door—but they came in a different door. I was sure it was for security purposes. I received a call on my cell phone, the secretary was coming in such and such a doorway. I hustled to get over there. We shook hands.

Secretary Powell said, "At your service, Wil, what do you want me to do?" First I took him into the Green Room. Five university presidents wanted to take photographs with him. We took those photographs and moved onto the main hall. I knew he was very efficient and didn't like wasting any time.

We walked in, and the place erupted in a wild ovation. I gave the welcome remarks and introduced Colin Powell, United States Secretary of State. A funny thing happened next. It was just the two of us on the platform. There was a pitcher of water and a glass to his right. After he had spoken maybe a minute to two minutes, he coughed. It got my attention. The second time after about three minutes, he coughed again. I thought, *Oh, maybe he has dry mouth*. He coughed a third time, and I got up to go around the table to get the pitcher.

He said, "Wil, just take a seat. At my first cough, you should have got me a glass of water. I'm going to have to talk to you about that."

The place roared with laughter. He spoke for about 45 minutes. He talked about his personal life and about the value of HBCUs and their many contributions to America. We walked him out surrounded by applause from the hall. Bottom line: we had a great visit. We had a similar situation with Condoleezza Rice, a very graceful lady. She came in and received the same reception.

At that time we were about to get a new secretary of education. Margaret Spellings was to replace Paige. I hated to see Paige go but that was not uncommon at the federal level. Secretaries come and go. People in all positions leave. Margaret Spelling came on board as the next secretary of education and I knew she would bring her staff, her people she wanted to give plum assignments to. I submitted my letter of resignation on November 30, 2005.

Shortly thereafter I received another call from a friend of mine who was in the U.S. Peace Corps, Jay Katzen, a former house of delegates member and a candidate for lieutenant governor. I had no idea he was over there. He knew I had left the U.S. Department of Education. He and his people talked to the White House and they wanted me to be the associate director for management at Peace Corps.

Director Gaddi Vasquez and Deputy Director Jody Olson, who subsequently became the director, interviewed me and hired me. As associate director for management, I was to handle two issues. The first issue had to do with personnel. On L Street, their headquarters in Washington had nearly 1,000 people on the staff. However, Peace Corps volunteers in 68 to 70 countries around the world were not getting the personnel support they needed. He wanted me to correct the situation. The second key issue was for me to find another location for the Peace Corps Headquarters.

Gaddi Vasquez was soon replaced by a new director from Minnesota, Ron Tschetter, and he swore me in. I was lucky again to have a great group of staff serving under me.

In that job, I needed assistance and support from the Peace Corps chief counsel because the litigation emanated when Peace Corps volunteers didn't get what they needed in the field, or when they left didn't get the benefits owed to them. This problem came under my area of responsibility.

My facilities director, Al Miller, was a retired Air Force chief master sergeant who was followed by another gentleman, Jim Pimpedly, also a retired Army NCO. They handled the facilities side.

We fixed the HR system, and we had the facilities side working so we were spending less money but giving greater service to the Peace Corps.

It was a 24/7 job. In case of inclement weather during the winter, I would get a call of closing or delayed openings; it was my responsibility to call into the headquarters recording to let the 1,000 people on staff know the federal government's decision. I had to keep my phone available, and I lived by it every night.

My rule of thumb was this: Less than eight inches of snow, I would come into work via train to Farragut West, and walk a block and a half to my office. Even if the employees were given a two-hour delay, I always went in to make sure the lights were on and all the entrances to the facility were cleared of snow. I required the same of my facilities director. If I was going to be there at 6 o'clock in the morning, I wanted him there, too. When our people come in, they would have safe, easy access to the building.

I retired after two years, leaving the Peace Corps stronger than it had been.

Epilogue

I believe that education is the key to success. Learning begins at home and goes all the way through life. Another key for me is strong interpersonal skills. I think I acquired those strong interpersonal skills from growing up in a family of six boys and one girl. I learned a lot about life and how to get along with others, right there in the house. My mother instilled in us the practice of treating each other with respect. Of course we had a few pushes and shoves, but that's to be expected.

From childhood I had a strong work ethic, which is required to be successful in this world. It was one of the earmarks of our family. Another was our love for each other, the support we provided each other, both with my siblings and then with Emily and our two children.

My wife is the most important person in my life, followed closely by my son and my daughter Lisa, even though she's no longer with us. They are the reasons I was so successful. Role models that I wanted for Lisa were two young ladies. First, Diane Hartley, a major in the military police corps, was just the ideal person that I would have wanted Lisa to emulate. Later, I met another young lady at one of the posts and I thought Lisa would have been like her. Col. Alexia Fields was the epitome of what an army officer should be like. Lisa's plan had been in the Army for her two-year obligation and then become an attorney. She had been interested in international law.

However, my son Wil and Lisa eclipsed all that I had done. That's all Emily and I ever wanted for our children. I think they took the best of both of us and excelled academically and in their careers. My son is my heart. My daughter is my heart. The bottom line is my partnership, the love and support of my wife Emily. I wouldn't have made it without her. During the time I served in Vietnam and on the DMZ in Korea and all the hardship tours, she always supported me, to give me the strength of mind to keep the focus on my job, so I could return home to them safely. I made the right decision to marry Emily as my lifelong partner and friend.

Acknowledgments

I am writing this book in the memory of my brother, Dr. Willie Lee Bryant, who was the writer in my family. Each year, our family celebrated the Thanksgiving holidays together and after the big meal, Willie would discuss his thoughts about writing our family history, "The Boys from the Fields." Willie never finished his book. In the fall of 2015, Willie was diagnosed with a terminal disease that resulted in his untimely death. Upon his passing, I made my promise to continue our story so that our children, family members, friends and colleagues will be able to keep our legacy alive for many generations. Therefore, as the reader journeys with me through this story, I will attempt to trace the struggles of how a determined, young single mother, Claudia Mae Bryant, was adamant on impressing upon her children that hard work, honesty and a good education were the keys to a better life in a fast, changing world.

By default, as the brother next to Willie, it was my responsibility to write this book. I am indebted to several people for the encouragement, wisdom and total support to accomplish my goal. I want to thank Julie Wakeman-Linn, retired English professor, writer, friend and neighbor, for all her help with creating this book.

I am deeply indebted to my wife, Emily, for her encouragement, advice and professional support. Reading has been, and continues to be, a passion for her since childhood; she provided positive guidance in seeing this project reach a finished product. Secondly, my son, Wil Jr., and my late daughter, Lisa, who always wanted their dad to capture our life experiences, for other family members, friends and the general public, to learn about some of the struggles, challenges and heartaches we endured, and to see how hard work, dedication and staying focused attributed to the numerous successes of our family.

I am also grateful to many who touched my life, influenced and encouraged me to write this book. First, my teachers from Mays High School: Carl Hanna, Mrs. Virginia Adger, Coach James "Jimmy" Anders, Coach Earl Dinkins and Coach Rufus Tribble.

At Florida A&M University, my thanks go to Director of Student Activities Rev. Moses General Miles, University Registrar E. M. Thorpe, as well as the outstanding men of Beta Nu Chapter, the Alpha Phi Alpha Fraternity and the ROTC Instructor Group in the Department of Military Science.

I'd like to thank Virginia governors George Allen and James S. Gilmore for the opportunities they gave me.

Most importantly, my mother: Claudia M. Bryant. A phenomenal woman, who taught me (and my siblings) that we could accomplish anything in life: with a good education, being respectful of self and others, a strong work ethic and being faithful to God's teachings.

Photo Captions

page 4	And inside back cover: The Bryant boys, around 1947. Standing, left to right: Willie Lee Bryant, Alvin Bryant, Wilbert Bryant; seated, Henry Louis Bryant, and Sylvester Bobby Bryant.
page 7	Claudia Mae Bryant, around 1938.
page 22	The Mays Rams Baseball team. Coach Dinkins on the right, Wilbert Bryant, kneeling, 2nd from right.
page 28	Wilbert Bryant, 1957.
page 36	Florida A&M ROTC. Wilbert Bryant, center front row, 1961.
page 47	ROTC summer camp, 1961 at Fort Benning, Georgia. Wilbert Bryant stands at ease, middle row, 4th from the right.
page 48	Top, Emily Mitchell in her white summer dress on the campus lawn of Florida A&M University. Bottom, Wilbert Bryant, dressed smartly for class as his fraternity required.
page 52	At their commissioning ceremony, ROTC cadets wore their tropical worsted tan Army uniforms. Wilbert Bryant is first on the right.
page 62	Wil and Emily, with their wedding party.
page 63	The newly weds prepare to cut the cake.
page 64	Wilbert and Emily, surrounded by family.

page 66   Lt. Wilbert Bryant serves as a briefing officer in the Joint Visitors Bureau for Swift Strike III headquarters in Spartanburg, South Carolina.

page 76   Lt. Wilbert Bryant gives another briefing in the Joint Visitors Bureau for Swift Strike III headquarters.

page 81   Above, Wil Jr. came into the world, kicking on January 13, 1965. Below, Wil Jr. was a cheerful and lively baby.

page 82   Father welcomes his six-month-old son to Germany in June, 1965.

page 87   In Germany, Wilbert Bryant is promoted to captain, with Emily pinning on one of his new insignia.

page 90   Wilbert Bryant, second from left, in Vietnam.

pages 100-101   Wilbert Bryant, second from right, in Vietnam, with his American and South Vietnamese advisory team.

page 110   At Fort Benning, Wilbert Bryant receives the Bronze Star for his service in Vietnam and is selected for promotion to major.

page 112   Korea 1969: As commander of troops, Bryant faces the general.

page 128   At Howard University in 1970, Commandant of Cadets Wilbert Bryant and the ROTC Faculty and academic team. Bryant, second from left, wears civilian dress as part of ROTC's official effort to defuse tensions of the era.

page 140   Wilbert Bryant and family visiting at the Defense Intelligence Agency, 1979.

page 146   The Bryants' ever-reliant VW bug.

page 150	Map of the United Kingdom and U.S. military bases thereon.
page 155	Wil Jr. and Lisa in Greece
page 161	Bryant, his mother, Claude, and Lisa in London, England
page 164	Lisa in second grade, 1978.
page 166	Colonel Bryant at the Pentagon.
page 169	The Bryants' home in Burke, and Lisa, striking a cheerleading pose, at the front door.
page 178	Battalion Commander Bryant, 1981.
page 182	Wil Jr. outfitted for football at Seaside High School.
page 187	Wilbert Bryant and his battalion in Panama, but no snakes in this picture.
page 199	Detail of watercolor of the National War College at Fort McNair, D.C., by artist Swabz, dated 5/27/05.
page 205	War College trip to Africa, visiting a village outside Harare, Zimbabwe.
page 213	As a freshman, Lisa Bryant is crowned Miss Robinson Secondary School.
page 224	And inside front cover: Col. Wilbert Bryant's retirement day, June 30, 1990.
page 227	Col. Wilbert Bryant speaks at his June 30 retirement ceremony.
page 237	Antique cars joined Virginia Union University's Homecoming Parade in Richmond, Virginia.
page 241	Family gathers to celebrate Lisa's graduation from Princeton.

page 243   Lisa, head of the cheerleading squad, at Princeton, stands atop a pyramid of cheerleaders.

page 253   Wil Jr., Lisa, and Emily enjoy Christmas with a real Christmas tree.

page 254   Secretary of Education Bryant speaks to students in the Capitol in Richmond, Virginia.

page 265   Deputy Secretary of Education Bryant visits Russia as part of a team from Virginia's Department of Education.

page 268   Virginia Secretary of Education Wibert Bryant meets U.S. Secretary of Education Rod Paige.

page 273   Virginia Secretary of Education Wilbert Bryant.

page 279:   Texas Gov. George W. Bush, poses with Virginia Secretary of Education Wilbert Bryant and his wife Emily Bryant, and Virginia Gov. Jim Gilmore, with his wife Roxane.

page 284   Wilbert Bryant wears the regalia of his first honorary doctorate, from Virginia State University.

page 286   The President's Board Advisors on Historically Black Colleges and Universities meets with President George W. Bush in the Oval Office.

page 289   Deputy Assistant Secretary Wilbert Bryant meets HBCU presidents and Pat Swaggert.

page 302:   Wil Jr., his children and mother on the day of his MBA graduation from University of Southern California.

page 304:   The Bryant family, Lisa and Emily, Wil Jr. and Wilbert.

Cover Photo Credits:

Wilbert Bryant, Deputy Assistant Secretary for the Office of Post Secondary Education. Public domain sourced/access rights from NB/DeptComm/Alamy Stock Photo 2K6704E https://www.alamy.com

Panoramic photo of a potato field on a sunny day. Copyright (c) 2021 Andrii Yalanskyi/Shutterstock. https://www.shutterstock.com 1904266585

"How does one measure a life? Perhaps a lively journey through the chain of command and interactions with star power of the military, Capitol Hill, and Peace Corps is enough. But for me it's the upbringing—raised by a single mother of six children, hiding from the KKK, off to Vietnam, then Korea, and the tour of HBCU's from Florida A&M to Howard University—that truly gives off sparks."

—Richard Peabody, editor, *Gargoyle Magazine*

www.ingramcontent.com/pod-product-compliance
Lightning Source LLC
La Vergne TN
LVHW061623070526
838199LV00078B/7407